MW01489669

ARCHAEC

Edited by
ERZSÉBET JEREM and WOLFGANG MEID

Series Minor
23

THE ARCHAEOLOGY
OF FIRE

Understanding Fire
as Material Culture

Edited by

Dragos Gheorghiu and George Nash

BUDAPEST 2007

Front Illustration:
Pit firing using dung as fuel. Vadastra village, August 2003
(Photo: Dragos Gheorghiu)

Back Cover Illustration:
Copper smelting in a pit. Archaeodromde de Bourgogne
(Photo: ABAB)

ISBN 978-963-8046-79-6

HU-ISSN 1216-6847

2007
ARCHAEOLINGUA ALAPÍTVÁNY
H-1250 Budapest, Úri u. 49

Copyediting by Réka Benczes
Desktop editing and layout by Rita Kovács

Printed by Prime Rate Kft.

Contents

List of Figures

Preface

This volume constructed from 12 carefully chosen chapters, specifically tackles fire and its presence within the archaeological record. The archaeological record suggests that fire plays a number of essential roles in society. Fire is a heatmaker, a homemaker, a lightmaker, a signalmaker and ritual-symbolic marker; it aids performance as well as serving the domestic and mundane. Despite these roles, fire within the archaeological record is seldom discussed. Within a ritual context, fire is considered merely as a tool of illumination. Within domestic contexts, fire is usually recorded as heating and cooking things. However, social anthropology and numerous ethnographies around the World tell a very different story. To many non-western societies fire is symbolically vital in everyday working life. Fire controls and manipulates the life cycle of many. It is not only a functional tool but it also possesses many supernatural qualities. Much oral story telling, from the Dream-time and to post-medieval poetry tells of mystical animals rising from or falling into fire. Social anthropologists such as Claude Lévi-Strauss talk of fire being part of structural opposition (i.e. Fire : Water). Ritual fire is also, according to the African ethnographic record, a vital tool in creating [ceramic] pots, not just as the heatmaker but also as heat injecting life and history into each vessel. One can assume that the fire power was also present in ancient pots. More importantly, the potmaker too would possess power. Both maker and the product would have constituted a special relationship that would have put them outside the mundane. The same can be said for the metalmaker.

It is clear that the secularity of present day society appears to take for granted of pyro-technology; one strike, one has heat and light. However, in ancient society, fire, similar to other day-to-day chaos, would have been approached in a number of very different ways. Again, according to LÉVI-STRAUSS (1986) there are essentially two types of fire. Both fires physically do the same thing. However, one is earthly, fulfilling the mundane needs of society, and the other is celestial and clearly used within ritual events. Based on a number of ethnographies fire transcends both worlds; from ignition to the disposal of the ash and embers (MOORE 1986).

In essence, this volume discusses a number of fundamental roles that fire plays. It is clear that fire is more than just fire and means different things to different people (PARKER PEARSON – RICHARDS 1994, 41). Papers within this volume, although diverse in content, follow three fundamental themes; archaeology, ethnography and experimentation. Chapters by Dods, Gheorghiu, Kroll-Lerner,

Nash, Odgaard and Purhonen focus on the sociology of fire, while Andrews, Frère-Sautot and Harding discuss the interaction between things, society and technology. Audouze and Rowlett discuss the problems of understanding the fragmentary evidence of the distant past; Rowlett boldly tackling the use of fire by early hominids.

The editors and contributors would like to thank the following people for their assistance. Firstly, many thanks to Ms. Julene Barnes and Dr. Jenny Moore for their initial input. Also thanks to Abby George (University of Bristol), to Becky Rossef for reading through various chapters and making essential comments, and to Cornelia Catuna for the final editing. Finally, thanks to Dr. Elisabeth Jerem and Ms. Réka Bences, of Archaeolingua Press in Budapest, for putting up with our usual excuses. All mistakes are the responsibility of the editors and the contributors.

October 2006   Dragos GHEORGHIU and George NASH

Bibliography

LÉVI-STRAUSS, C. 1986
  The Raw and the Cooked, Harmondsworth: Penguin.

MOORE, H. L. 1986
  Space, Text and Gender: An Anthropological Study of the Marakwet of Kenya, Cambridge: Cambridge University Press.

PARKER PEARSON, M. – RICHARDS, C. 1994
  Architecture and Order: Spatial Representation and Archaeology. In: Parker Pearson, M. – Richards, C. (eds.), *Architecture and Order. Approaches to Social Space*. London & New York, Routledge, 38–72.

Introduction

Firemaker!

DRAGOS GHEORGHIU – GEORGE NASH

The management of fire is a subject as vast as the history of humankind and, paradoxically, a minor subject for the archaeological literature of the last decades of the 20th century (for books published on this topic see PERLÈS 1977; WERTIME – WERTIME 1982; GOUDSBLOM 1992; CANNON – CONNOR 1993; COLLINA-GIRARD 1998; SCOTT – MOORE – BRAYSHAY 2000; REHDER 2000; PYNE 2001; GHEORGHIU 2002; FRÈRE-SAUTOT 2003). Consequently, the purpose of the present book, a result of a session organized at the Lisbon EAA Meeting in 2000 (GHEORGHIU 2000, 120–128), would be to sensitize archaeologists in the new millennium to perceive fire as the subject of a special archaeology, not only as a phenomenon but as a material agent (*cf.* HOUGH 1926), alone, or in association with material objects, a subject widely discussed in the papers forming this book.

Evidence of early fire

Comparable to the development of humanness, due to some morphological changes due to a diet based on cooking plant food (see WRANGHAM *et al.* 1999), territorial expansion on the earth's surface is also the result of fire management. This allowed the colonization of temperate areas as well as the expansion into many zones with a permanent harsh climate as for example the sub-Arctic Canada and Greenland, where fire determined and centered all human activity (ODGAARD, this volume). Fire allowed life in the glacial Europe and Eurasia as well as the colonization of Australian continent, where one could presume human presence some 55 kyr, which coincided with the extension of a large number of gigantic animals, probably due to the firing of the land.

If at the middle of the 20th century DART's (1925, 1948) *Australopithecus Prometheus* revealed to have been an immature hope to demonstrate a high antiquity of fire management (see OAKLEY 1956). Recent discoveries from the sites of Swartkrans (South Africa) (1.5/6 myr), Koobi Fora (Kenya) (1.5 myr) or Chesowanya (Kenya) (1.5 myr) confirmed that even if the use of fire did not begin simultaneously with the fabrication of stone utensils, which is dated circa

2.5 myr in East Africa (HARRIS – CAPALDO 1993), it has, nonetheless, a very ancient history.

Some of the oldest fireplaces in the world archaeologically identified are from Koobi Fora, east of Lake Turkana, belonging to *Homo erectus/ergaster*. They are an index of a major structural change of the psyche, i.e. a surpassing of the instinctual animal's fear of fire and the transformation of the humanoid into a pyro-philic and pyro-dependent being. At Koobi Fora the traces on the soil were not produced by natural fires (as inferred by JAMES 1996, 71) but by a long time cultivated fire with temperatures around 400° C, as the study of phytolits demonstrate (ROWLETT, this volume), being a sort of predecessor of the hunter-gatherer "central places". Fireplaces were places where hominids returned to light the fire from time to time, their role being of heat, light and protection against predators (BELLOMO 1994), and served also for the preparation of food.

Most recently, at the meeting of the Palaeontology Society, Brian Ludwig's study of the thermal alteration of the stone tools excavated from contemporary sites with Koobi Fora showed that the potlid fractures on the stone tools, post 1.6 myr, could be the index of the emergence of a pyro-culture at that time. Experiments revealed that circa 1.3 myr ago at Swartkrans site in South Africa (SILLEN – BRAIN 1990) bones were fired at 800° C, a temperature higher than that of grass fires which only desiccate-cook organic materials at 200° C. Other experiments were made to determine the use of fire at Koobi Fora (BELLOMO 1994).

In the history of early fire there is a still unexplained large gap of time between these first proofs of fire management and the later frequent presence of fire as ashes, cinders or burned bones and stone dated circa 460,000 years ago in Asia at Zhoukoudian (China), or in Europe at l'Escale (France) or Vértesszőlős (Hungary) (DOBOSI 1990), which are solid date in favor for a extended management of fire.

The advent of pyro-centric cultures

Ice Age in Europe and Eurasia compelled the two co-habitant human species, Neanderthal and Modern man, to use fire for countless operations, as the cooking of food, the preservation of food by smoking, the hardening of the wooden weapons, the transformation of raw materials as flint, ochre, bones, etc. Fire provided the necessary heat and light (POTTIER 2002, 19), and was the centre of a socializing space; the complexity of the fireplace was an indicator of the type of occupation of a site. For example at Abri Pataud (France) early Aurignacian I phase, the distribution of hearths (MOVIUS 1966) expresses the spatial pattern of a temporary camp (BINFORD 1988, 163) that used fire to process animals, to prepare tools and colors and to rest (POTTIER 2002, 24). One can also add that fire had symbolic and supernatural qualities as well, used to illuminate the back walls, niches and overhangs of caves. Cave artists would have applied their pigments to rock panels to paint supernatural beasts involved in hunting scenes. In order to construct, say, the large frescos at the cave sites of Chauvet (Ardèche) or Peche Merle (Dordogne), light from hearths and touches would have played an essential role, not just for the artists but for those who dared venture into the depths of cave to witness these beasts.

During the Gravettian one can notice a new the utilization of fire for baking clay figurines at almost 800° C (SOFFER – VANDIVER 1994, 1997), as at Pavlov where 4000 wet clay figurines were fired in a hearth sheltered by a windbreak made of mammoth bones coated with clay. Some other 6000 fired clay figurines were fired at Dolni Vestonice I in two proto-kilns consisting of pits with a partially domed enclosure (JOCHIM 2002, 75) which denotes an advanced control of fire, but also a short duration of use for the pyro-instruments.

In the Magdalenian Upper Palaeolithic, as for example at the open site Verberie in the northern Paris Basin, in a hunting camp of short duration, related to the fall reindeer migration (AUDOUZE 1987; AUDOUZE – ENLOE 1991), one can note along the centering of all human activity around the hearts the emergence of a specialization in the early use of fire. One hypothesis, inferred from ethnological models, could have been a specialization of the fires, one of the hearts being employed to smoke the meat (AUDOUZE, this volume). Palaeolithic fires were fuelled with defleshed bones, jointly with wood, and, certainly, with the animal fat, as archaeological and ethnological models show (see MARCH 1996, 258).

A mobility of fire analogous with that identified in the Palaeolithic (WATTEZ 1996, 31) could be found only at the historical nomadic populations (KROLL-LERNER, this volume), sometimes the material culture of these peoples being used as model for understanding prehistoric archaeological record.

The sedentariness of fire begun with the Neolithic, when it was confined to permanent places, as the ceramic container of the oven, inside the house, and the control of combustion was performed by controlling the air currents.

Starting with the Neolithic ceramics began to be the skeuomorphic plastic material *par excellence*, being fired in low depressions in the soil, in pits, in ovens or draught kilns (KINGERY 1997; GHEORGHIU 2002, 83 ff). From now on ceramic pyro-objects as domed ovens (PREVOST-DERMARKAR 2003, 221), fixed or portable kilns, braziers, ambers protectors, mobile altars and systems of illumination were common to every settlement and are the index of sedentariness as well as of a specialization of the domestic fire. As soon as the Late Neolithic, a new instrument designed to control fire, the draught kiln, created a ceramic mass production and a craft specialization (GHEORGHIU, this volume).

In the ages to come, the control of the utilitarian fire would be improved by the sedentary populations, to cite only the production of fine ceramics in oxidized and reduced atmosphere (ANDREWS, this volume), the casting of metals or the fabrication of glass.

The study of the prehistoric archaeological record infers the existence of an anthropology based on the management of fire which involves identity, sacredness, rituality, anthropomorphism, in relationship with fire. For example, in Neolithic, and in the later epochs across Europe, one could identify a continuity of meaning and identity of the dwelling space by means of the continuation of the household fire (HARDING, this volume), as well as a separation between a feminine pyro-oikos and a masculine pyro-agrios. Beginning with the Neolithic an alternative funerary custom of transforming the human body by means of cremation is identified archaeologically, this ritual with powerful signification being in use, with interruptions, until the present days (PURHONEN, this volume).

A further case of Neolithic ritual firing was the burning of houses (TRINGHAM 1992; STEVANOVIC 2002) and of the settlements at the moment of abandon, the Neolithic/ Chalcolithic tells of South Eastern Europe being a sort of multi-layered macro objects formed by the overlapping of cyclical fired horizons. The fire ecology of tells seems to have been analogous with that of the cultivated fields, burned to be fertilized, being at the same time a method of fixing the event in the material and in memory, a method of purification and of consolidation of the foundation ground for the following overlapped levels of construction.

This custom of firing the abandoned settlement was continued in the Bronze Age of South Eastern Europe. It seems that Bronze Age favored fire, because of the multitude of domestic ceramic pyro-objects and pyro-landscapes, as the rows of cooking pits from Northern Europe (THÖRN, this volume). In the pyro-landscape fire was employed also to animate burial activity within (and may be outside) the chambered monuments of Europe.

Luminosity would have played a vital role in the performances associated with burying the dead, in particular in late Neolithic passage grave traditions of NW Europe. Archaeological evidence points towards both hearths and torches being employed inside both the chamber and passage areas of many monuments. The luminosity and its affect would have allowed both the living and the dead to witness the architecture of the tomb. Much of the megalithic art associated with the passage grave tradition would have played a vital role in the performance of internment (NASH, this volume).

Metals which replaced ceramics as social prestige in the Bronze Age, conferred to the craftsman a special status, as described by Hesiodus (*Works and Days*, verses 105 to 202), due to the mastery of a complex pyro-technology; kilns were replaced by high temperature furnaces necessary to melt and separate the ore. Copper exploitation was possible owing to the heat fracturing effect produced by firing large quantities of wood against the mine's walls (O'BRIEN 1996, 22).

As soon as the Bronze Age, and mainly in the Iron Age, fire is archaeologically identifiable in the social conflicts and is actively participating to the shaping of architecture, urbanism and environment. With the emergence of cities urban fire becomes a reality that lasted up to our days.

Representations of fire

The weak representation of fire in prehistoric and protohistoric visual arts was due to its phenomenological character that is difficult to be visualized iconically. Even in the realist ancient Greek art the representations of fire were very rare, and generally not depicting the process of combustion, but only the moment of pre-ignition. Fire was preferably represented indirectly since only its material supports, as altars, lamps, pyres, ritual places, etc., or its material avatars, as fire gods were depicted.

Such a visual avoidance of the phenomenon, in all prehistoric and proto-historic art, could mean that fire was probably "better to think" than better to be represented. All Palaeolithic cave art was dependent of fire which animated the painted images in the darkness of the caves, so it could have been *associated to*; or it could have been *the creator of* these images. In the Middle to Upper Palaeolithic art the animation of paintings with torches or oil lamps is obvious, being a sort of transfer of vital force from the lighting source to the image (*anima* in Latin means *soul*), as at the Chauvet cave (see MOHEN 2002), where animals were figured with many legs or were represented in a sequence of movements like the filmic images. The bison with 6 legs positioned in different attitudes, the deer with 5 heads as well as the row of felines preparing to attack the bison would create the effect of a life-like motion when the animator/performer would have moved the source of light on the surface of the painting.

Another hypothesis is that fire could have been a component of the painting process and the Palaeolithic colour chart of red and yellow ochre, black manganese and grey-white from the European caves paintings could have represented the colours of lit and of extinguished fire (figured as ashes and charcoals), this also suggesting the existing of a direct symbolic relationship between art and fire. In this perspective colors probably signified the life and death of the painted animals, analogous with the process of burning and the quenching of fire.

The good fire or the anthropology of pre-modern man

There are a number of analogies between fire and humans, to cite the limited lifespan, the heat of the human body, the colour of the blood, sexuality, the mystical states; this is why, for example, the unpredicted character of fire is to be found under the anthropomorphic shape of the trickster fire gods.

Indo-European gods as Agni/Hermes (for a relationship between Agnis and Hermes see HOCART 1970; BERG 2000) or Hestia, could have been the expression of gender in relationship with fire, and this symbolism could infer the existence of two kinds of gendered fires.

Fire was determinant for the anthropology of the pre-modern man: the perception was developed for a total sensorial reception of the phenomenon, as temperature, colors, sounds, and smells; the daily life (work and sleep) took place in the proximity of thermal areas, and the rites of passage were marked by it, as the solstice or the funeral fires. The fire of the hearth was central to daily life like the heart is to the body, as suggested by archaeological record and by the etymology of these words in English, and the sacredness of the phenomenon was transferred to the pyro-instruments (i.e. the material supports as hearths or altars) which contained it. Sacredness could have been found also in the processes of kindling and quenching the fire and in the way the combustion remains were ritually deposited or recycled, and could have originated from the heavenly origin of fire, from the process of magical extraction from materials, or because of the capacity to be a go-between the world of men and deities/ancestors, this conferring to fire the character of an inter-realm communicator.

To demonstrate the importance of fire for the pre-modern man, we chose three of the 1028 hymns of the *Rig-Veda*, probably the richest collection of hymns dedicated to Agni, the "God Fire" (ELIADE 1977, 282, n.1):

« *I praise Agni, domestic priest, divine minister of sacrifice,*
Invoker, greatest bestower of wealth.
Worthy is Agni to be praised by living as by ancient seers;
He shall bring hitherward the gods.
To thee, dispeller of the night, O Agni, day by day with prayer,
Bringing thee reverence, we come;
Ruler of sacrifices, guard of the Law eternal, radiant one,
Increasing in thine own abode.
Be to us easy of approach, even as a father to his son;
Agni, be with us for our weal. (I, 1, 1–2, 7–9)

Thou, Agni, shining in the glory through the days, art brought
to life from out the waters, from the stone;
From out the forest trees and herbs that grow on ground, thou,
Sovereign lord of men, art generated pure.
Thine is the Herald's task and Cleanser's duly timed;
Leader art thou, and Kindler for the pious man.
Thou art Director, thou the ministering priest; thou
Art the Brahman, lord and master in our home.
Agni, men seek thee as a father with prayers, win thee,
bright-formed, to brotherhood with holy act.
Thou art a son to him who duly worship thee, and as a trusty
friend thou guardest from attack.
By thee, O Agni, all the immortal guileless gods eat with thy
Mouth the oblation that is offered them.
By thee do mortal men give sweetness to their drink.
Pure art thou born, the embryo of the plants of earth. (II, 1, 1–2, 9, 14)

Shine forth at night and morn: through thee with
Fires are we provided well.
Thou, rich in heroes, art our friend.
Bright, purifier, meet for praise,
Immortal with refulgent glow,
Agni drives Rakshasas away.
Agni, preserve us from distress;
Consume our enemies, O God,
Eternal, with thy hottest flames.
And, irresistible, be thou a mighty iron fort to us,

With hundred walls for man's defense,(...)
Infallible! By day and night. (VII, 15, 4, 8, 10, 13–15)

(Translation by Ralph T. H. Griffith, 1889–91, *The Hymns of the Rigveda*,
I–III, Benares, adapted by M. Eliade)

Pyro-cultural traits

When analyzing the large variety of instances of the use of fire in prehistory
and protohistory some patterns become recurrent, and their identification helps to
understand the anthropology of pre-modern man:

Fire was essential to human life, being a strong socializing factor between the
living, between the living and the deities and between the living and the dead or
ancestors.

It was the generator of speech and, later, of history.

It provided protection, light and warmth.

It was utilized for food preparation and food conservation.

It helped to conquer the night.

It divided the domestic area into different zones.

It had a double role, simultaneously private and public.

It had a character of separation, of liminality and simultaneously of connection
between different realms, being a medium of passage, and being connected with
the rites of passage.

It was a symbol of signaling.

It was used for sacrifices.

It was in relationship with the living body and with the corpse.

It was used for modeling Nature, as landscape or materials.

It was used to transform the characteristics of materials.

It had the character of shaping the thermoplastic materials.

It had the character of fixing things and structures after being burned.

It was used in social life as well as in social conflicts.

It was used as a purifying element.

It regenerated vegetation and fertilized and purified the cultivated fields.

It was a generator of gender construction, by engendering the domestic space.

It divided the daily time as well as the cosmic time.

It had a cyclical character.

It animated images in the dark.

The evil fire or the anthropology of Modern man

One of the consequences of the « progressive devaluation of the cosmos » (DODDS 1965, 37), in progress with Late Antiquity, was the transformation of fire into a punitive instrument and the decline of its importance in daily life. One can infer that this process of accusation, in progress from Late Antiquity, produced a "scotomization of fire", by transforming it into a negative phenomenon, related to punishment and disaster. With the process of industrialization, fire took on new associations, being perceived in almost all the cases as an "accident", or risk factor. After the Industrial Era humans were no longer dependent on the domestic fire, which became invisible (see PYNE 2001, 115) being hidden in the insides of the machine.

Today we are sensitized to the role of fire in society by the evil consequences of the wild fires of wars, natural disasters and accidents. Almost everyday, mass media show that behind contemporary civilization could stand a pyro-technic disaster. Probably this absence of fire from the anthropology of contemporary man could explain the current absence of an archaeology of fire.

Towards a pyro-archaeology

There is no archaeologist that has not been confronted with the indexes of fire as material culture, during the excavations, but until today, material artifacts are more important to archaeological research than the process of making and controlling fire. Behind almost every object or organization of living and funerary spaces hides a pyro-technology that is waiting to be discovered, reason why the purpose of the present book is to endorse a pyro-archaeology that will promote a complex study of fire as an archaeological discipline. A first step was made with the present volume by perceiving fire as a material culture element.

The role of a pyro-archaeology would be the study of fire from a micro to a macro level through the direct analysis of archaeological remains and through the indirect method of experimentation, and the discipline will connect antracology, the pyro-technologies of transforming the nature of materials as ceramic studies, archaeometallurgy, glassmaking studies, the pyro-technologies of building and destroying things, the pyro-technologies of cremation, techniques of food preparation and conservation, systems of heating, techniques of landscape modeling, techniques of war. Last, but not least, studies on the symbolism and phenomenology will complete the agenda.

Bibliography

AUDOUZE, F. 1987

The Paris Basin in Magdalenian times. In: Soffer, O. (ed.), *The Pleistocene World*, Plenum, New York, 183–200.

AUDOUZE, F. – ENLOE, J. 1991

Subsistence strategies in the Magdalenian of the Paris Basin. In: Barton, N., Roberts, D. – Roe, A. (eds.), *The Late Glacial in North-West Europe*, CBA Research Report 77, Council for British Archaeology, Oxford, 63–71.

BELLOMO, R. V. 1994

Methods of determining early hominid behavioral activities associated with the controlled use of fire at FxJj 20 Main, Koobi Fora, Kenya, *Journal of Human Evolution* 27, 173–95.

BERG, P.-L. 2000

Hermes & Agni: A fire-god in Greece?, *Journal of Indo-European Studies* 40, 189–204.

BINFORD, L. R. 1988

In Pursuit of the Past. Decoding the archaeological record, Thames and Hudson.

CANNON, K. – CONNOR, M. 1993 (eds.)

The Society for American Archaeology fire symposium 1990, *Archaeology in Montana*, vol. 23.

COLLINA-GIRARD, J. 1998

Le Feu avant les allumettes, Edition de la Maison des Sciences de l'Homme, Paris.

DART, S. A. 1925

A note on Makapansgat: a site of early human occupation, *South Africa Journal of Science* 22, 371–51.

DART, S. A. 1948

The Makapansgat proto-human Australopithecus Prometheus, *American Journal of Physical Anthropology* 6, 259–83.

DOBOSI, V. T. 1990
Fireplaces of the Settlement. In: Kretzoi, M. – Dobosi, V. T. (eds.), *Vértesszőlős: Site, man and Culture*, 519–521.

DODDS, E. R. 1965
Pagan and Christian in an Age of Anxiety, Cambridge, Cambridge University Press.

ELIADE, M. 1977
From Primitives to Zen. A Thematic sourcebook of the history of religions, San Francisco, Harper and Row.

FRÈRE-SAUTOT, M.-C. 2003 (ed.)
Le Feu domestique et ses structures au Neolithique et aux Ages des metaux, Monique Mergoil, Montagnac.

GHEORGHIU, D. 2000
The Archaeology of Fire, *Book of Abstracts*, 6th EAA, Lisbon, 120–128.

GHEORGHIU, D. 2002
Fire and air-draught: Experimenting the Chalcolithic pyroinstruments. In: Gheorghiu, D. (ed.), *Fire in Archaeology*, BAR 1089, 83–94.

GOUDSBLOM, J. 1992
Fire and civilization, Harmondsworth, Penguin.

HARRIS, J. W. K. – CAPALDO, D. S. 1993
The Earliest Stone Tools: Their Implications for an Understanding of the Activities and Behavior of late Pliocene Hominids. In: Berthelet, A. – Chavaillon, J. (eds.), *The Use of Tools by Human and Non-Human Primates*, Oxford, Oxford University Press, 196–220.

HOCART, A. M. 1970
Kings and Councillors. An Essay in the comparative anatomy of human society, Chicago/London, University of Chicago Press.

HOUGH, W. 1926
Fire as an agent in human culture, Smithsonian Institute United States National Science Museum, bull. 139.

JAMES, S. R. 1996

Early hominid use of fire: Recent approaches and methods for evaluation of the evidence. In: Bar-Josef, O., Cavalli-Sforza, L. L., March, R. J. – Piperno, M. (eds.), *The Lower and Middle Palaeolithic*, ABACO Edizioni, 65–76.

JOCHIM, M. 2002

The Upper Palaeolithic. In: Milisauskas, S. (ed.), *European Prehistory. A Survey*, New York, Boston, Dordrecht, London, Moskow, Kluwer Academic/Plenum Publishers, 55–113.

KINGERY, W. D. 1997

Operational Principles of Ceramic Kilns. In: Rice, P. (ed.), The Prehistory and history of ceramic kilns, *The American Ceramic Society* VII, 21–40.

MARCH, R. J. 1996

L'Étude des structures préhistoriques: Une approche interdisciplinaire. In: Bar-Josef, O., Cavalli-Sforza, L. L., March, R. J. – Piperno, M. (eds.), *The Lower and Middle Palaeolithic*, ABACO Edizioni, 251–275.

MOHEN, J.-P. 2002

Arts et préhistoire. Naissance mythique de l'humanité, Paris, Terrail.

MOVIUS, H. L. 1966

The Hearts of the Upper Perigordian and Aurignacian horizons at the Abri Pataud, Les Eyzies, Dordogne, and their possible significance, *American Anthropologist*, Vol. 68, No.2, part 2, 296–325.

OAKLEY, K. P. 1956

The Earliest fire makers, *Antiquity* 30, 102–107

O'BRIEN, W. 1996

Bronze Age Copper Mining in Britain and Ireland, Dyfed, CIT Printing Services.

PERLÈS, C. 1977

Préhistorie du feu, Paris, Masson.

POTTIER, C. 2002

Un habitat du gravettien moyen a burins de Noailles de l'Abri Pataud. In: Gheorghiu, D. (ed.), *Fire in archaeology*, BAR International Series 1089, 19–26.

PRÉVOST-DERMARKAR, S. 2003

Les Fours Néolithiques de Dikili-Tash (Macédoine, Grèce): une approche expérimentale des téchniques de construction des voutes en terre à batir. In: Frère-Sautot, M-C., (ed.), *Le Feu domestique et ses structures au Néolithique et aux Ages des métaux*, Montagnac, Monique Mergoil, 215–223.

PYNE, S. J. 2001

Fire. A brief history, London, The British Museum Press.

REHDER, J. H. 2000

The mastery and uses of fire in antiquity, Montreal, Kinston, London, Ithaca, McGill-Queen University Press.

SCOTT, A. – MOORE, J. – BRAYSHAY, B. 2000 (eds.)

Fire in the Palaeoenvironment. Special issue of *Palaeogeography, climatology and ecology* 164.

SILLEN, A. – BRAIN, C. K. 1990

Old flame: Burned Bones provide Evidence of an Early Human Use of Fire, *Natural History* 90 (4), 6–10.

SOFFER, O. – VANDIVER, P. 1994

The Ceramics. In: Svoboda, J. (ed.), *Pavlov I: Excavations 1952–3*, ERAUL 66, Lige, 161–173.

SOFFER, O. – VANDIVER, P. 1997

The Ceramics from Pavlov I1957 Excavations. In: Svoboda, J. (ed.), *The Dolni Vestonice Studies*, Vol. 4, Academy of Sciences of the Czech Republic, Institute of Archaeology, Brno, 383–401.

STEVANOVIC, M. 2002

Burned houses in the Neolithic of South East Europe. In: Gheorghiu, D. (ed.), *Fire in Archaeology*, BAR International Series 1089, 55–62.

TRINGHAM, R. 1992

Households with faces. The Challenge of gender in prehistoric architectural remains. In: Gero, J. – Conkey, M. W. (eds.), *Engendering archaeology. Women in prehistory*, Blackwell, Oxford and Cambridge, 57–92.

WATTEZ, J. 1996

Modes de formation des structures de combustion: Approche méthodologique et implications archéologiques. In: Bar-Josef, O., Cavalli-Sforza, L. L.,

March, R. J. – Piperno, M. (eds.), *The Lower and Middle Palaeolithic*, Forli, ABACO, 29–34.

WERTIME, T. A. – WERTIME, S. F. 1982 (eds.)
Early Pyro-Technology: The Evolution of the first Fire-Using Industries, Washington D.C., Smithsonian Institution Press.

WRANGHAM, R. W. – HOLLAND JONES, J. – LADEN, G. – PILBEAM, D. – CONKLIN-BRITTAIN, N. L. 1999
The Raw and the Stolen. Cooking and the Ecology of Human Origins, *Current Anthropology* 40, 5, 567–594.

1

Between Material Culture and Phenomenology: The Archaeology of a Chalcolithic Fire-powered Machine

DRAGOS GHEORGHIU

Methodology

The methodological problems of *an archaeology of fire* depend on the approach to the subject, since fire is at the same time material and phenomenon; therefore, in order to cover the complexity of the subject, the study of fire deserves a joint approach, combining the two aspects of the topic.

Consequently, the present paper focused on the study of fire used in the Chalcolithic up-draught kilns will try to cover the technological aspects of material culture that shapes fire, as well as the phenomenological aspects of the process of combustion. For the study of material culture I will begin with a technological analysis of the functionality of kiln's forms, and discuss the construction and functioning through the perspective of experiments. As experiment is at the same time an ethic and emic approach, in order to demonstrate the complexity of the subject, I will discuss, in a final paragraph, the emic, sensorial, experience of fire achieved in the course of experimentation.

The control of fire in Neolithic South Eastern Europe

An examination of Neolithic pyrotechnology reveals that the firing of objects was performed on the soil surface (in bonfires or using ceramic containers, as the ovens, DESHAYES 1974; COMSA 1976; COMSA 1996; PREVOST-DERMARKER 2003; VITELLI 1997), and in the soil, in depressions and pits, and had in common the following problems:

Low temperatures (for bonfires see WOODS 1982, 18–19; GIBSON – WOODS 1997, 212; GOSSELAIN – SMITH 1995; CARLTON 2002, 70; MARTINEAU – PETREQUIN 2002, 345 ff; GOSSELAIN 2002, 169, fig. 144; for pits and vase-kilns see GHEORGHIU 2002a, 86); uneven temperature during combustion; fire clouds (see GIBSON – WOODS 1997, 156) and specific colours on the ceramic objects produced by a direct contact of the fired objects with the fuel; the breakage of a part of the objects due to spalling (see GIBSON – WOODS 1997, 252) as the result of a direct contact with the flame the impracticability of controlling the flow of cold air in open fires, therefore the risk of sudden contraction of hot ceramics and the formation of cracks in the fired material; the use of different fillers in the paste to absorb thermal shock. These had generally a large granulation and did not allow the making of objects with very thin walls, or later, the use of these pastes on the potter wheel. An incomplete firing of the core of the clay objects; a large diversity of colours resulted in a single pit firing (experiments showed that in a single fire there was possible to produce both oxidized and reduced atmosphere, with a very large diversity of nuances on the objects fired) (GHEORGHIU 2002a).

Consequently, a new pyrotechnology was needed to produce a higher, even temperature, and to protect the clay objects during the firing process and during the process of cooling, and this was possible with the help of a new instrument, the draught kiln.

Straining and sieving within Chalcolithic pyrotechnology

In my opinion the development of Chalcolithic pyrotechnology is the result of the second products revolution (SHERRATT 1981), because there seems to have been technological analogies between the food and firing processes.

The food operations used the methods of sieving or straining for the processing of vegetal or of dairy products; such inference is supported by archaeological remains, as the ceramic vases with perforated bottoms often found in the Lower Danube region from the 5th millennium BC Archaeologists assigned all objects with perforated walls to the category of "cheese strainers", even if their form seemed to have been designed for a different function, like the objects with perforated corbelled walls, that experiments demonstrated to function as «Bunsen lamps» (see WOOD 1999, GHEORGHIU 2002a), or in my interpretation, after recent experiments, as "amber protectors" and "fire starters". Such functions related more to pyrotechnology than to food production, demonstrate that perforations could have had the role of fire control, the perforated walls allowing air draught, and this new perspective on Chalcolthic design would reclassify many objects as being fire starters or braziers (see GHEORGHIU 2002a; GHEORGHIU 2003a, 41, fig. 5).

The Chalcolithic air-draught kiln –
the first machine functioning with fire

Draught kilns are archaeologically attested from the 6th millennium BC in Mesopotamia (see SIMSON 1997, 39) and in the South Eastern Europe around the 5th–4th millennia BC (see MARKEVICI 1981 and COMSA 1976, 25), where they were utilized along with bonfires and pit fires. Until today in the archaeological record there is evidence only for under ground draught kilns, but this does not exclude the existence of surface draught kilns, which, as experiments demonstrate, resisted only few seasons to weathering and human agency.

The coming out of draught kilns corresponds to a specialization of the functions in prehistoric object design, and this object, the result of the exploitation of perforated surfaces with air-absorbing properties, could be perceived as the first machine that functioned with the use of fire as energy.

One of the principal reasons of the emergence of the draught kiln was the firing of a large number of clay objects at once, and the intention for a mass-production led in time to the emergence of a craft specialization and to a particular axiology of ceramics. In Chalcolithic South East Europe it seems that ceramics have had an important social role which could be inferred from the amount of energy spent for forming and decorating clay objects as vases, figurines, as well as from their relationship in mortuary contexts where they are found amongst other items that are considered to have represented "wealth". The up-draught kiln functions on the principle of the control (see RYE – EVANS 1976, 164; ARNOLD 1997, 213) of the ascending flow of hot gases and flames.

Major functions of the draught kiln are heat containment (SHEPARD 1956, 75; KINGERY 1997, 11) and heat transfer to wares, as well as the protection of ceramic objects inside the firing chamber from thermal shocks when heated or cooled. Like for the open fires, any organic fuel could be used (see ARNOLD 1997, 213), from dung and household residues to reed and wood.

An important function that differentiates the draught kiln from the other prehistoric pyro-instruments is the gradual evacuation of water from the molecular structure of the clay, during the process of combustion, since the controlled elimination of water does not produce spalling, a typical accident for uncontrolled combustion in bonfires (see GIBSON – WOODS 1997, 252). Other functions are the protection of the fire process against humidity (rain or moisture) (ARNOLD 1997, 213) and the production of higher temperatures (SHEPARD 1956, 75) compared to the other pyroinstruments already mentioned, which led to the production of a high quality ceramics.

The description of the functioning of the draught kiln

Draught kilns could be divided into two categories (downdraft and updraft kilns) depending on the direction of the flow of hot gases; in the present text I will discuss only the up-draught type ("the simplest form of kiln", HAMER 1975, 306), frequent in the archaeological record. The up-draught kiln used in experiments was copied after those discovered in the final phase of Cucuteni-Tripolye tradition (5th–4th millennia BC) at Costesti (MARKEVICI 1981) and Glavanestii Vechi (COMSA 1976, 25).

This complex pyro-object is composed of several forms with the following functions (*Figure 1.1*):

Figure 1.1. Cross-section through a corbelled up-draught kiln showing the draught of hot gases and flames through the fire tunnel, the perforated platform and the firing chamber (Photograph D. Gheorghiu).

- A fire tunnel/firebox in front of which the fire is lighted, and gradually introduced inside;
- A firing chamber, shaped as a cylindrical or corbelled container with an upper opening/vent;
- A perforated platform positioned at the base of the firing chamber (*Figure 1.2*). Some kilns had a border to fix the perimeter of the platform on it (see COMSA 1987, 99);
- A platform support. The shape of the supports found in the archaeological record are cylindrical or rectangular, in the latter case with perforations made at the base to allow a homogenous circulation of the hot gases in the interior of the combustion chamber (the up-draught kiln from Glavanestii Vechi, Cucuteni-Trypolye tradition); and
- A system of covering the vent which allows the air-draught.

Compared to the other pyroobjects, the up-draught kiln offers a series of advantages:

Figure 1.2. The perforated platform with a thermocouple on its surface (Photograph D. Gheorghiu).

- a separation of the fuel from objects, and subsequently a separation of the objects from the direct flame;
- a separation of the firing space into a combustion chamber (at the base) and an upper firing chamber, by means of a perforated platform;
- a straining of the hot gases and flames through the platform's perforations (in reality a set of tubes for air absorption), which produces a distribution of the thermal shock;

Figure 1.3. Sealing the vent of the kiln with shards covered with wet clay.
The operator is fixing a thermocouple (Photograph D. Gheorghiu).

- the handling of the fuel during the process of firing with the effect of regulating the interior temperature through the positioning of the fuel in the firing tunnel and the opening or closing of the vent. By sealing the openings of the kiln (*Figure 1.3*), and subsequently by consuming the oxygen, one can create a reduced atmosphere that will turn out the colour of fired ceramics into black tones (*Figure 1.4*), due to the loss of oxygen by the iron oxides contained in the clay (see GIBSON – WOODS 1997, 235). Such advantage of creating both oxidized and reduced atmosphere during a single process of firing by means of air control led to a special method of firing (analogous to the three-stages firing performed in Antiquity) in Neolithic South Eastern Europe (see VITELLI 1997, 30).

Figure 1.4. Vases fired in a reduced atmosphere, seen from the vent (Photograph D. Gheorghiu).

The control of air flow and subsequently of fire by regulating the time and temperature of firing by means of manipulating the fuel and by closing the tunnel and vent apertures (RYE – EVANS 1976, 164) confers to the up-draught kiln the quality of a "machine", i.e. a complex instrument that could be regulated, having at the same time a certain autonomy in the process of functioning.

Chaîne-opératoire of construction. Description after experiments[1]

Although starting as a pit dug in the soil, after repeated firings the up-draught kiln becomes a ceramic object like the oven or the brazier, this is why I believe that the kiln, as a material support of, and as a result of fire, is the best example for the syntagma "fire as material culture" from the subtitle of the present book. For archaeologists the result of this process of transformation is to be found currently in the archaeological record, where the fired interior surface of the kilns looks like the walls of a large ceramic vase (*Figure 1.5*).

Figure 1.5. The opening of a load fired in an oxidized atmosphere.
Note the ceramic circle formed around the vent (Photograph D. Gheorghiu).

---

[1] The experiments of building and firing up-draught kilns were conducted in Vadastra village (south of Romania); with the collaboration of Dr. Alex Gibson between 2000 and 2002, and individually, between 2001 and 2004. A total of 13 firings (4 in surface kilns and 9 in underground kilns) were performed (see GHEORGHIU 2002b; GHEORGHIU 2003b). The 2002 experiment was financed by a Romanian Ministry of Culture Grant, the 2001–2002 experiments by a CNCSIS-World Bank Grant, the 2003 experiment by a CNCSIS grant.

The efficient shape of the kiln was due to the small dimensions of the openings, designed to preserve heat, although these un-ergonomic dimensions made difficult the loading and unloading of the firing chamber. However, experiments demonstrate that some of the up-draught kiln shapes were dimensioned according to ergonomic factors. Therefore the height of the kiln allowed the operator to work at the level of the firing tunnel and, at the same time, to observe the surface of the vent, to see the release of steam (*Figure 1.6*) and flames, reason why often kilns were positioned on low terraces, with gentle slopes which required little human intervention. Although ethnographic studies show a propensity of modern potters for locating the kilns on positions exposed to wind (see ARNOLD 1997, 218), in the experiments I carried out, the best location was in areas protected from the wind blows, as the lower river terraces.

The experiments suggested that the *chaîne-opératoire* of the construction of an underground kiln could have been the following:

Figure 1.6. The kiln seen from the stoke pit.
One can notice the strong elimination of steam (Photograph D. Gheorghiu).

- The digging of a 2 m deep corbelled pit (with diameter of 1.50 m at the base and 0.50 m at the upper part/the vent);
- The digging of the firebox/fire tunnel (1 m long with 0.40 m diameter);
- The finishing of the walls of the corbelled space with a rough layer of clay mixed with straws which had to be pressed and hammered with a piece of wood, and afterwards overlapped by a fine layer of clay mixed with dung;
- The construction of the foot/support of the platform. Experiments demonstrated that the most efficient method was to preserve the material of the support when digging the corbelled combustion chamber;
- The making of the clay platform on a twigs structure. The construction could be made entirely in the interior of the kiln, or by putting together parts of the platform modeled outside the kiln, and the perforation to be performed from outside, through the vent; and
- The natural drying of the kiln by means of natural ventilation followed by a first firing to remove the water from the fabric of the clay walls.

Chaîne-opératoire of firing.
Description after experiments and ethnographic studies[2]

Ethnographic studies from literature (HAMER 1975; DECA 1982) and from the field, carried between 2000 and 2002 in the north of Oltenia, as well as the experiments carried in Vadastra revealed an obvious physical determinism in the *chaîne-opératoire* of firing, which depends of the chemical processes occurring in the clay fabric during the combustion process which, due to its repetitive character, could suggest a ritual action when using the kiln fire: the loading of the firing chamber through the vent. Due to the small aperture of this opening, and of the low resistance of the perforated platform, the operation could have been carried by children (*Figure 1.7*). After the loading, the vent was covered with ceramic shards to protect the interior and at the same time to allow air absorption through the ceramic strain created in this way; the positioning of the fuel in front of the fire tunnel in such a manner as to allow air draught through the pieces of fuel. Every firing starts with a stage of heating the kiln, with the heated air passing through the tunnel, the combustion chamber, the perforated platform, the load of ceramic

---

[2] The ethnographic studies were carried on in the villages of Oboga and Romana, North of Oltenia County. For the first experiments I benefited of the help of Mr. Adrian Tambrea, a potter from Horezu.

objects and, finally, through the covering of the vent. The efficiency of kilns depends on the degree of dryness of their walls and platforms, and experiments with a kiln with freshly plastered walls conducted to low temperatures, in spite of the large quantity of fuel used. A diagram of such a firing showed rhythmical rises and downs in the process of augmenting the temperature, revealing the function of the kiln as being a sort of pump that rhythmically extracts the water from its component parts, until the complete drying of the clay.

Figure 1.7. Loading the kiln with the help of children (Photograph D. Gheorghiu).

Comparable low temperatures occur when using wet fuel, so a simple method is to place the wet fuel over the vent (see ARNOLD 1997, 218), to be heated by the hot air evacuated. A kiln that was fired recently and is dry and relatively warm has a better efficiency, compared to its first firing. In an up-draught kiln the draught of hot gases and flames is so powerful that radiated heat in the walls is nearly absent, this phenomenon being observed in surface up-draught kilns whose walls become warm only after the closing of the vent and the cessation of the air-draught; the observance of the first stage of firing. The rising of the temperature is a slow process, especially at the beginning of the combustion, imposing a stage between 20° and 120° C (see HAMER 1975, 22) to allow water smoking, i.e. the elimination of water from the pores of the air-dried clay; the time spent at this stage is in relationship with the dimension of objects and the thickness of the walls of the vases fired. Up to 350° C, the rapid decomposition of organic material from the clay's fabric is signalled by a specific smoke; the observance of a second stage of firing. Between 350° to 700° C, the process of ceramic change occurring in the clay causes the sintering of the internal particles (KINGERY 1997, 12). This process imposes a stage at 570–600° C to allow the elimination of zeolitic water, a process going together with an emission of black smoke, a specific smell and the beginning of an emission of red light from the clay objects. From 700° C on, the emission of light increases as soot accumulates on the lower surface of the shards covering the vent (*Figure 1.8*); at the level of 800° C, the flames from the combustion chamber reach the vent surface. However, as experiments demonstrate, the vent of the kiln could be protected against weathering with wooden structures covered with inflammable vegetal materials without the risk of fire-starting, if they are placed at a distance of 1.50 m height from the shards cover. The highest temperature reached in the up-draught kiln built in Vadastra was under 1000° C, due to the dimensions of the fire tunnel which gradually filled up with embers and ashes thus diminishing the air flow, so to go over this temperature the fire tunnel was cleaned up and a new fire was started. As a consequence, the temperature lowered by 150° C and afterwards jumped over the threshold of 1000° C. This operation was the most difficult to be performed because of the high temperature radiated when extracting the embers with wooden instruments, which had to be executed in a very short interval of time, together with the restarting of the fire before the temperature in the firing chamber lowered too much. The radiated heat around 1000° C is perceived by the pyrotechnologist as a sort of invisible solid wall in front of the working area.

Vitreous traces occurred on the lower surface of the platform when temperature went over 1050° C, followed by the total plugging of the tubes of the platform at 1150°–1200° C and the transformation of the colour and fabric of the objects close to the platform. When during an experiment the temperature in the combustion chamber went over 1200° C, the platform built from alluvial clay collapsed because all its air draught tubes were filled with glass and slag, which suggests the use of refractory materials for this part of the kiln. There are differences between the techniques of firing concerning the duration of the process. Ethnographic examples suggest a long time for firing (between 12 and 18 hours, see DECA 1982, 214), but successful experiments were performed in less than 8 hours, when the temperature at the middle of the load was around 900° C, and the ceramic objects became translucent. The apertures of the kiln are sealed up, to allow a homogenization of the interior heat, and, later, a slow cooling of the ceramic load. The close up is for minimum 8 hours, and could be made by sealing up the two apertures with large shards and wet clay; the unloading of the kiln could be executed by at least two persons: one individual will descend head down into the firing chamber to collect the vases since the other/s will grasp his feet. The following tasks would be undertaken:

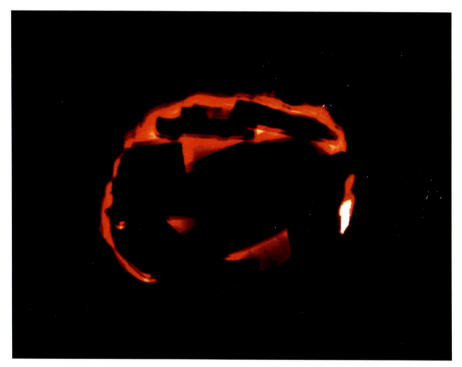

Figure 1.8. The shard cover of the vent, seen after the emission of soot (Photograph D. Gheorghiu).

- The cleaning of the firing chamber from the fired clay that fell when unloading the kiln;
- The cleaning of the perforations of the platform which had the tendency to fill up;
- The cleaning of the ashes that accumulate in the fire tunnel after every fire;
- The periodical cleaning of the wasters;

- The periodical/seasonal (?) repair of the firing chamber by plastering the cracked walls;
- The protection of the kiln during the wet season with earth and shards; experiments demonstrate that surface kilns could resist to weathering only for a few seasons before turning to dust; in contrast, a subterranean kiln with the vent protected during the winter resisted successfully for many campaigns; and
- As instrument for controlling the fire one could use only a stick with a length over 1.80 m and with one end arched.

The sensorial experimentation of fire

The description of the *chaîne-opératoire* of firing revealed the up draught kiln as being a machine that shapes the fire, regulates it and is created by it, and whose technological phases which are standardized because of the physical constraints in the material (as water elimination or sintering) could be perceived as technological – ritualistic actions, in a direct relationship with the human senses because the colours of the flames or of the fired objects, as well as the smells and the radiated heat, inform about the interior process of combustion. Since all technological operations had a correspondence in the sensorial area one can observe the utilization of the machine through a phenomenological perspective:

- The water elimination could be perceived visually and olfactory, the emission of steam being followed by a specific smell;
- The firing stages could be perceived through the senses too: visually (a blue range of colours between 300° to 500° C and a red range between 600° to 900° C and an orange-yellow range between 900° to 1150° C) and thermal. I noticed that, starting with the temperatures from red heat range the kiln created thermal spaces of varied intensities that could be perceived from small distance with the help of the thermal detectors in the skin and by olfactory perception; and
- The whole process of combustion involves the perception of time, expressed in the logic of the succession of operations related to the sensorial perception and the quantity of fuel consumed.

It seems probable that all operations of firing took place during the night time from two determinant reasons: the calm atmosphere of the night allowed a better visual, olfactory and thermal control of the fire, as well as a steady air-draught. (*Figure 1.9*)

*Figure 1.9. The pyrotechnologist looking to the fire tunnel
before extracting the embers at 900° C (Photograph D. Gheorghiu).*

Concluding remarks: The magic of the fire machine

When analyzing the south-east European Chalcolithic material culture, one can observe that the "second products revolution" signified at the same time a change (or specialization) in objects' design as well as a change at the social level. A comprehensible example for the design change is the up-draught kiln (the first machine with multiple separated functions, in which fire is locked and regulated by means of straining, like the alimentary products, and by a sensorial control), and for the social change is the emergence of a new craft specialization and, as a consequence, a mass production of high quality ceramics.

By analyzing the technological improvements in ceramic technology in the large context of the emergence of the Chalcolithic stratified society, one can infer a special status for the new specialized craft of the serial production of fired clay objects by using a machine to regulate fire.

But it is only the emic experience of the *chaîne-opératoire* that allowed me to deeply understand the complexity of the up-draught kiln and to re-evaluate this machine, as being one with a total embodiment of the technology and a strict rituality of operations, and, in this way, to reconsider the role of the pyrotechnologist in Chalcolithic society. I believe that the control of the machine (i.e. the controlled transformation of matter), the increase of the quality and quantity of the production of one individual, as well as the nocturnal activity of the kiln conferred to the pyrotechnologist a special prestige, probably with a magic role. I will never forget my first experience when I fired a kiln without the help of thermocouples in a summer night in the village of Vadastra; at that moment I realized that my body influenced, and was influenced at the same time by the fire, and this special psychical and kinaesthetic relationship with fire revealed to me the magic power of controlling the element as well as being controlled by it.

Acknowledgements

I would like to express my thanks to all who helped me in developing the pyro-experiments with kilns in the Vadastra village: Dr. Alex Gibson (University of Bradford), Mr. Adrian Tambrea (Horezu village), Mr. Ion Cococi, Mr. Marin Batranca (Vadastra village). Last, but not least, my gratitude to the art and design students (National University of Arts in Bucharest) and to the Vadastra village children who contributed to the project with the construction of vases.

Bibliography

ARNOLD, E. D. 1997
    Ceramic theory and cultural process, Cambridge University Press, Cambridge, New York, Port Chester, Melbourne, Sydney.

CARLTON, R. 2002
    Some comments on the technology of prehistoric pottery in the Western Balkans in the light of the ethnoarchaeological research. In: Gheorghiu, D. (ed.), *Fire in Archaeology*, BAR International Series 1089, 63–81.

COMSA, E. 1976
Caracteristicile si insemnatatea cuptoarelor de ars din aria culturii Cucuteni-Ariusd, *SCIVA* 27, 1, 23–33.

COMSA, E. 1987
Neoliticul pe teritoriul Romaniei. Consideratii, Bucharest, Editura Academiei RSR.

COMSA, E. 1996
Vetrele si cuptoarele din locuintele neolitice din Muntenia, *Istorie si traditie in spatiul romanesc,* III, Bucharest.

DECA, E. 1982
Centrul de ceramica din comuna Lungesti, Jud. Valcea, *Buridava – Studii si materiale*, Ramnicu Valcea, Ramnicu Valcea Museum, 211–216.

DESHAYES, J. 1974
Fours néolithiques de Dikili Tash*, Mélanges helléniques à Georges Daux*, 66–91.

GHEORGHIU, D. 2002a
Fire and air-draught: Experimenting the Chalcolithic pyroinstruments. In: Gheorghiu, D. (ed.), *Fire in Archaeology*, BAR International Series 1089, 83–94.

GHEORGHIU, D. 2002b
The Vadastra Project: Experiments with traditional technologies, *Old Potter's Almanach*, vol. 10, no.1, March, 9–10.

GHEORGHIU, D. 2003a
Water, tells and textures: A multiscalar approach to Gumelnita hydrostrategies. In: Gheorghiu, D. (ed.), *Chalcolithic and Early Bronze Age Hydrostrategies*, British Archaeological Reports International Series 1123.

GHEORGHIU, D. 2003b
Archaeology and Community: News from the Vadastra Project, *Old Potter's Almanach*, vol. 11, no.3, June, 1–4.

GIBSON, A. – WOODS, A. 1997
Prehistoric Pottery for the archaeologist, London and Washington, Leicester University Press.

GOSSELAIN, O. P. 2002
Poteries du Cameroun méridional. Styles techniques et rapports à l'identité, Paris, CNRS Editions.

GOSSELAIN, O. P. – SMITH, A. L. 1995
The Ceramic and Society Project: An Ethnographic and Experimental Approach to Technological Choices, *KVHAA Konferenser*, 34, 147–160.

HAMER, F. 1975
The Potter's Dictionary of materials ad technologies, London, Pitman Publishing; New York, Watson Guptill Publications.

KINGERY, W. D. 1997
Operational Principles of Ceramic kilns. In: Rice, M (ed.), *The Prehistory and History of Ceramic Kilns*, Proceedings of the Prehistory and History of Ceramic Kilns, 98th Annual Meeting of the American Ceramic Society in Indianapolis, 11–20.

MARKEVIC, V. I. 1981
Pozne-Tripolskie plemena severnog moldovii, Kishinev.

MARTINEAU, R. – PÉTREQUIN, P. 2002
La Cuisson des potteries néolithiques de Chalain (Jura), approche expérimentale et analyse archéologique. In: Pétrequin, P., Fluzin, P., Thiriot J. – Benoit, P. (eds.), *Arts de feu et productions artisanales*, Antibes, Éditions APDCA.

PREVOST-DERMARKER, S. 2003
Les fours Néolithiques de Dikili Tash (Macédoine, Grèce). Une approche expérimentale des techniques de construction des voutes en terre à batir. In: Frère-Sautot, M.-C. (ed.), *Le feu domestique et ses structures au Néolithique et aux Ages des Métaux*, Montagnac, Ed. Monique Margoil.

RYE, O. S. – EVANS, C. 1976
Traditional pottery techniques of Pakistan: Fields and laboratory studies. Smithsonian Contributions to Anthropology, No. 21.

SHEPARD, A. O. 1956
Ceramics for the archaeologist. Carnegie Institution of Washington, Publication 609.

SHERRATT, A. 1981

Plough and pastoralism: aspects of the secondary products revolution. In: Renfrew, C. – Shennan, S. (eds.), *Ranking, resource and exchange.* Cambridge, Cambridge University Press, 13–26.

SIMSON, J. 1997

Prehistoric ceramics in Mesopotamia. In: Freestone, I. – Gaimster, D. (eds.), *Pottery in the making. World Ceramic Traditions.* London, British Museum Press.

VITELLI, K. 1997

Inferring firing procedures from shards: Early Greek kilns. In: Rice, P. M. (ed.), *The Prehistory and History of Ceramic Kilns*, Proceedings of the Prehistory and History of Ceramic Kilns, 98th Annual Meeting of the American Ceramic Society in Indianapolis, 21–40.

WOOD, J. 1999

Bunsen burners or cheese mould? A new reinterpretation of a Bronze Age ceramic. *Abstracts*, 5th EAA Meeting, Bournemouth.

WOODS, A. 1982

Smoke gets in your eyes: Patterns, variables and temperatures in open firings, *Bulletin of the experimental firing group*, vol.1, 11–25.

2

Hearth and Oven in Early Iron Age Sobiejuchy, Central Poland

ANTHONY HARDING

The antiquity of the hearth as centrepiece of house and home is well known. Many ancient peoples are known to have regarded the hearth as crucial both functionally and symbolically to the continuation of life. The reasons are of course not far to seek: the hearth provides warmth, it is the place where meals are cooked and maybe served, it provides a certain amount of light; it is natural that both family and visitors might seek out its welcoming presence in the cool of the night, and especially in winter, to eat round, maybe to sleep near.

Traditions often dictated that the fire on the hearth must never be allowed to go out, as known most typically from Greek and Roman life and manifested in most obvious form in the "common" or "public" hearth in the *prytaneion* of Greek cities, or the eternal flame guarded by the Vestal virgins at Rome. Colonists leaving the mother city took with them the lighted flame to their new foundation and preserved it there. The fire on the hearth thus symbolised the permanence of city, house and home, for those who lived there. To let the flame die out would be tantamount to allowing the city to die. So the hearth came to be regarded as undying, immortal, in fact a goddess – for the Greeks, Hestia, for the Romans, Vesta. This was just one element in the Indo-European situation, where from Vedic India through to Celtic Ireland fire and the hearth are said to have had a special significance (MARINGER 1976). ANGELA DELLA VOLPE (1990) also explored these matters, pointing out that the domestic hearth was the symbol of the basic social nucleus, and through the rites performed around it, the tangible expression of the group's religious ideology. Interestingly for our present concerns, she goes on to maintain that the cellular social structure which is represented by the practice of tying each social unit to its homestead, or the soil it occupied, required the separation of one hearth from another, hence one household from the next; and ultimately to the custom of setting formal boundaries between people and households, through various intermediate stages connected with the ceremonial activities conducted at the hearth, and the importance of ancestor worship.

There are much later instances, in myth and history, where the hearth assumes a central importance in peoples' lives. "Heilig ist mein Herd; heilig sei Dir mein Haus!", as Hunding said to Siegmund in Richard Wagner's version of the myth (*Die Walküre*, Act I), but he was reflecting the traditional knowledge that Germanic societies also regarded the hearth as sacred, embodying and symbolizing the family, the continuance of life, and source of well-being, comfort, and shelter (HUTH 1939).

We know well where the large, public hearths were situated in Greek and Roman cities. We also know quite a lot about the prehistoric hearths that preceded these Iron Age formal hearths. While Minoan Crete is usually regarded as having made little use of permanent fixed hearths, using portable tripod hearths instead (HUTCHINSON 1962, 227; GRAHAM 1962, 137, 215), mainland Greece used megaron buildings from the Neolithic on, and a central hearth was the norm in the main room of such buildings – best known from Mycenaean palaces but also present in many other sites and contexts (e.g. VERMEULE 1964, 176).

This is not the place to embark on a full review of the significance and symbolism of the hearth in later prehistoric Europe; suffice it to say that there are plenty of cases where hearths that look to be special in some way are present. The famous example at the Wietenberg (Dealul Turcului) near Sighişoara in central Transylvania is a case in point (HOREDT – SERAPHIN 1971, 74 ff., Abb. 59). Cristian Schuster and Traian Popa have recently listed hearths on Bronze Age sites in Romania (SCHUSTER *et al.* 2001, 31 ff.), attributing a cultic function to this hearth (*ibid.* 40 f.); but most hearths in houses of the earlier Bronze Age in Romania, and on tell sites in Hungary, were simpler clay platforms, many times renewed. In Hungary, the early part of the Bronze Age saw the construction of both flat and framed hearths (*Rand-* or *Kesselherde*), usually round but sometimes elliptical or rectangular, as known above all at Tószeg (BANNER *et al.* 1959, 68 ff.), though this does not seem to have been duplicated on more recent excavations at other sites in Hungary (BRONZEZEIT IN UNGARN 1992). The Wietenberg hearth looks to be a striking forerunner of those in classical cities such as Seuthopolis in Bulgaria (DIMITROV 1961, 98, Fig. 4, Pl. XI (a)).

The cultic importance of fireplaces reappears at various times and places in the Bronze Age world. Particularly notable are the *Brandopferplätze* of the Alpine area, typically located in elevated positions and containing large amounts of burnt material (KRÄMER 1966; GLEIRSCHER 1996). These sites are, however, quite different in character from hearths in domestic contexts, and probably served a quite different purpose. As such, they will not be considered further here.

Numerous ethnographic studies have confirmed that such a picture is far from unusual in societies through history and through the world. JANET CARSTEN's (1997) study of a fishing community in Malaysia stresses that the houses never have more than one hearth and the hearth or *dapur* gives its name both to the room in which it is situated and to the rooms immediately adjacent to it. These rooms are seen as extensions of the cooking stove and are closely associated with women. The hearth in this community is central to living: even if co-residential families do not eat together, what was important is that their food is cooked on the same hearth. The hearth is the heart of the house, and cooking on it and eating at it are so crucial that there is actually a prohibition on eating in another house. The hearth has symbolic functions: it transforms raw materials into edible substance that becomes bodily substance; it produced blood and regulates its flow, and thus ensures reproduction, so it is intimately connected with childbirth practices. By its transformations of these raw materials, it converts things from outside the house into life-giving substance.

A fascinating study by GILL (1987) of attempts to introduce more energy-efficient stoves into various developing countries in the 1980s found that there was considerable resistance to them, for various reasons: speed of cooking, versatility, production of space heat, and the fact that traditional fireplaces served as a social focus and had a symbolic value being those most often quoted as the advantages of the traditional hearth. To quote Gill, "in Ghana, the 3-stone fireplace symbolizes a united family, whilst in parts of Nepal, villagers believe that a spirit dwells in their traditional hearth". They were used for heat, light and smoke, even the latter serving a purpose for smoking foodstuffs and repelling insects.

So much for background, which I quote in order to show that there really does appear to be a strong cross-cultural element to the centrality and importance of the hearth. So when we move to the Early Iron Age of Poland, the site of Biskupin naturally attracts most attention, and here there are the well-known streets with houses closely packed on either side (KOSTRZEWSKI 1950). The layout of the houses (*Figure 2.1*) followed a highly regular pattern, with an anteroom or vestibule, a main inner room on the right, with hearth in its centre, and an area for storage and perhaps animal stalling on the left. The hearth consisted of a circular area of stones, with the timbering of the floor stopping at their edge. Presumably the stones formed a base on which the fire was lit, and served as a means of refracting heat much like a modern fire-brick. It is easy to imagine that the position of the hearth gave it significance very similar to those of the Greek and Roman examples.

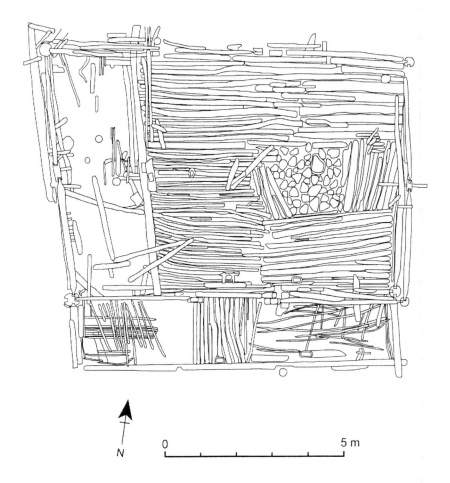

Figure 2.1. Plan of a Biskupin house. Source: KOSTRZEWSKI 1950.

Sobiejuchy is another large fortified site of the Early Iron Age in central Poland (*Figure 2.2*). It is close both geographically and culturally to Biskupin, from which it lies about 16 km distant. The site extends over about 6 ha and was originally, like Biskupin, an island separated from the adjacent mainland by perhaps 50–100 m of open water. Around it ran a series of defensive installations, consisting of a rampart, palisades, and breakwaters. Excavation in the interior produced extensive remains of domestic character, though it was not possible by this means alone to define house walls with any certainty. For this, we turned to geophysical survey, and the use of a Geoscan gradiometer in 1988 enabled us to produce this plan, subsequently processed using the Insite programme (*Figure 2.3*). On it you can see the lines of streets, running irregularly across the site, and beside the streets, the clear outlines of individual houses. The dark

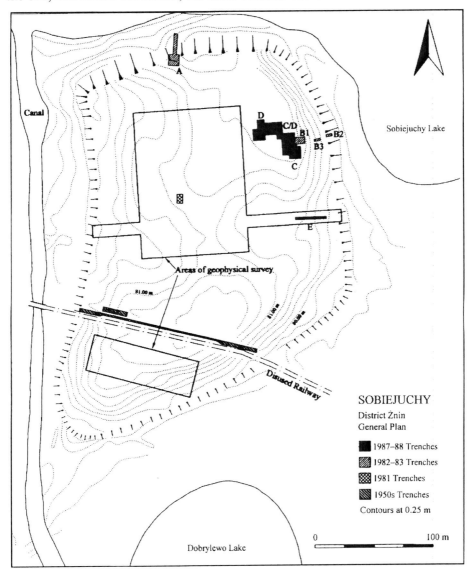

Figure 2.2. Sobiejuchy, woj. Bydgoszcz, general plan.

features may be in part lines of daub, and in part pits, hearths, or ovens. It is hard to be certain which of these are complete building plans, or what their dimensions are, but the largest may be around 22 m long and 5 m wide; others might be 12 x 7 m or thereabouts. The streets are about 5 m wide. In total there are probably around 20 houses visible here, though some are sketchily represented. You will note that there are also some areas where there appear to be no buildings; these were open areas, and remained so throughout the life of the settlement (which remains uncertain, though it all seems to fall within the Ha C period and perhaps lasted no more than 50–100 years).

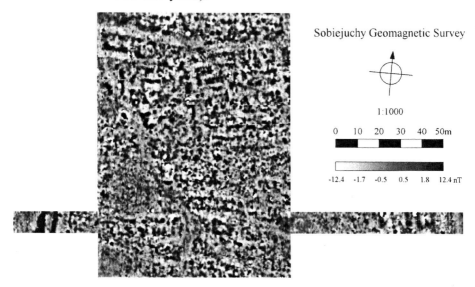

Figure 2.3. Geophysical survey plan of the central area of Sobiejuchy.

So we have here a large-scale site, divided up rather irregularly into streets and houses, and this of course reminds us that Biskupin was divided up much more regularly, though there are still sizeable parts of the site that have never been investigated. At Sobiejuchy a couple of installations used stones, usually smaller than those at Biskupin, but more examples were what we interpreted as ovens, with a flattish baked clay surface, usually crackled and orange from heat. Polish colleagues found a good example in the northern part of the site (*Figure 2.4*). We assume that these oven floors were surmounted by a dome-like clay cupola, and served for baking bread or other dishes. As happens with simple bread ovens at the present day, a wood or charcoal fire would be lit on the surface, and allowed to burn until the whole thing was hot, then the coals would be pushed

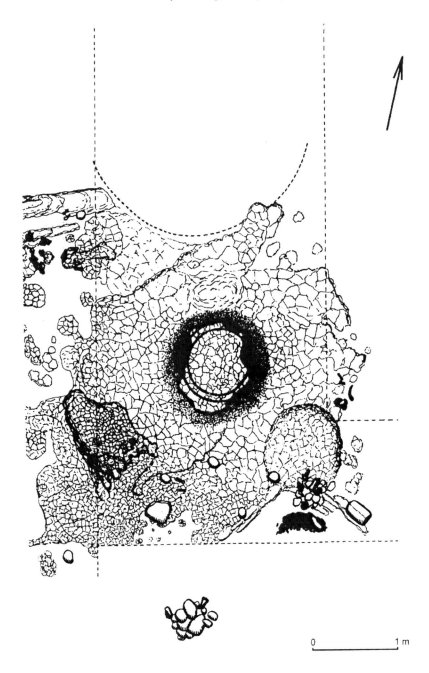

0   1 m

Figure 2.4. Oven surface from the Polish excavations
in the north part of the Sobiejuchy site.

Anthony Harding

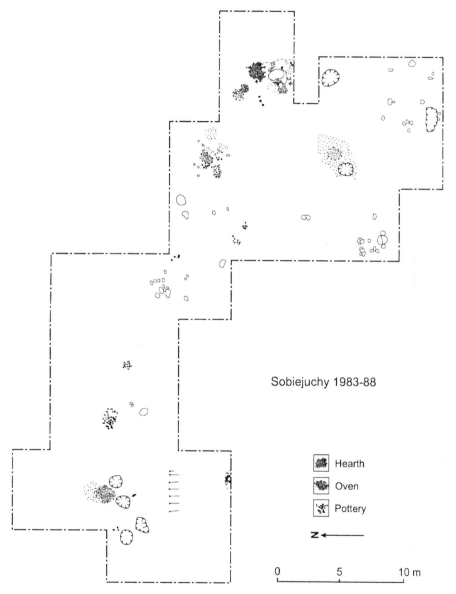

Figure 2.5. Plan of the excavated areas at Sobiejuchy showing hearths and ovens.

to the sides and the bread inserted. If a house served as the residence of a single-family unit, you would expect that each would have an oven, and conversely, that one oven would serve one house. When we look at the distribution of ovens at Sobiejuchy, we can see that there is some truth in this assumption (*Figure 2.5*). In

Area D, a large oven is surrounded by pits, pottery deposits, loom-weights, and other features; east of this is an area with much charcoal and sizeable amounts of animal bone; this may have been the cooking hearth. In Areas B and C, by contrast, the situation is more complex: although we believe we are looking at the interior of a single house here, there are no less than three ovens and a hearth, the latter cobbled, and with storage pits close by it. Now of course one possibility is that we are in fact looking at more than one house here, though this could not be recognised; or that we are looking at chronologically distinct ovens, that moved over time as houses were rebuilt on slightly different spots. Again, this is not something that we were able to recognise in excavation.

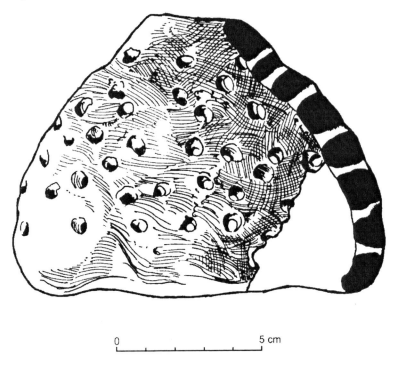

0   5 cm

Figure 2.6. Portable oven from Sobiejuchy.

A number of other finds show an intimate connection with fire and its use. Firstly, portable ovens (*Figure 2.6*) – there are a number of fragments of these, including one complete profile that unfortunately came from a trench cut through the edge of the site to investigate the environmental history rather than the cultural layers. But these objects are well-known, especially (though not only) in the Lausitz culture area. They are quite small, and so cannot have been used for major cooking enterprises. It is generally thought that would have been used to cover a hot coal taken from the fire, which would give out heat and possibly scented smoke; they could have been used as plate-warmers, or more likely as cup-warmers: the small hemispherical cups that were found in abundance on the site would sit nicely on the top, though anything larger would topple over. But these objects are perhaps not all they seem. Louis Nebelsick has recently explored (COBLENZ – NEBELSICK 1997, 28 ff.), in the context of the Lausitz cemetery at Niederkaina, the common practice of placing the "oven models" in graves, which are always in the group of material placed away from the cremation urn. This suggests a connection between the creation and maintenance of fire in life and the need to provide heat and food, in symbolic form, for the dead after life.

These finds, taken in conjunction with the presence of cooking and heating installations in the Sobiejuchy houses, suggest that special significance attached to sources of fire and the impedimenta that surrounded them. Yet it is curious that at Biskupin it is the hearth that was the centrepiece of the house, while at Sobiejuchy it was the oven. Does this tell us something about the way each was conceived and used?

For someone living in a Biskupin house, the hearth, with its glowing embers, smoke, maybe occasionally sparks (though for obvious reasons these were undesirable in a wood and thatch building), formed the natural pivot on which life turned. Once inside the house, you could not miss it. It was visible, audible, sensible to nose, eyes and ears; provided warmth; was the place were food was cooked if not actually prepared. At Sobiejuchy, on the other hand, hearths appear to have been less important and placed in no one special place, and instead the over assumed greater significance.

Sometimes the hearth must have been the main source of light as well as warmth, since little is known of lamps here or at Sobiejuchy. Interestingly, though we are talking about fire here, and the hearth is the home of fire, we are talking about a very restrained fire – no leaping flames and dancing shadows, or the house would soon be no more – unless the roof openings were unusually large – and given the proximity of one house to another, the whole settlement would soon follow suit.

If these historical and ethnographic examples provide some sort of confirmation of the importance of the hearth, and by implication the fire, in various societies, what of the oven? And what of the other, smaller-scale equipment that accompanied the making and moving of fire? The enclosed oven is generally regarded as having arrived on the scene relatively late by comparison with the open hearth, and I know of no specific cult activity associated with it in early historic Europe. But maybe we are barking up the wrong tree here. Maybe it was precisely because of the danger of open fires that the oven was created and used – coals might have been brought from open fires kept in the open air, and placed in the oven. By using a bellows, they would be kept hot, and by continuing the process the entire oven would heat up and act like a stove, being usable for baking as well – though not, of course, for boiling liquids for which an open flame really was essential.

I suggest that at least some of these installations at Sobiejuchy served a purpose as much symbolic as functional. The presence of pits around some of the ovens suggests that they served a central role in their buildings, but others are isolated.

Support for this idea comes from another puzzling feature. Deposits of carbonised grain were found close by the eastern hearth, some in a pot that had been placed in a pit, others in discrete patches beside the hearth. Radiocarbon dates obtained on these showed that two belong as expected to the mid- first millennium BC, but surprisingly, two others are dated to a much earlier period, around 500 years earlier, though there is no or hardly any stratigraphic distinction observable. Luminescence dating confirms that two phases are represented. At first I thought these were normal storage pits or pots, but it is tempting to believe they were deposits carefully placed and their positions in the ground carefully guarded over ensuing centuries, again emphasising the centrality and longevity of hearth and oven.

What we may be witnessing in these houses at Sobiejuchy is the specific process of creating domestic ritual, and following therefrom the boundaries between social units. Although more obvious at Biskupin, where the separation between houses is fully developed and each hearth and therefore social unit is completely isolated from its neighbours, at Sobiejuchy there was a much more fluid situation, which perhaps reflects the fact that it certainly started life much earlier, far back in the Bronze Age. If I am even half right, there are profound implications present here for the development of both social and religious life in later prehistoric Europe, and these are down to the humble – or not so humble – hearth and oven.

Since this article was submitted, the final publication of the Sobiejuchy site has appeared: Harding, A., Ostoja-Zagorski, J., Palmer, C. and Rackham, J. 2004. Sobiejuchy, a fortified site of the Early Iron Age in Poland. Warsaw: Institute of Archaeology and Ethnology of the Polish Academy of Sciences.

Bibliography

BANNER, J. – BÓNA, I. – MÁRTON, L. 1959

Die Ausgrabungen von L. Márton in Tószeg, *Acta Archaeologica Hungarica* 10, 1–140.

BRONZEZEIT IN UNGARN. 1992

Bronzezeit in Ungarn. Forschungen in Tell-Siedlungen an Donau und Theiss, Frankfurt am Main, Museum für Vor- und Frühgeschichte.

CARSTEN, J. 1997

The Heat of the Hearth: the process of kinship in a Malay fishing community, Oxford, Clarendon Press.

COBLENZ, W. – NEBELSICK, L. 1997

Das prähistorische Gräberfeld von Niederkaina bei Bautzen, Band 2. Veröffentlichungen des Landesamtes für Archäologie mit Landesmuseum für Vorgeschichte Dresden, Band 25. Stuttgart, Konrad Theiss.

DELLA VOLPE, A. 1990

From the hearth to the creation of boundaries, *Journal of Indo-European Studies* 18 (1–2), 157–84.

DIMITROV, D. P. 1961

Seuthopolis, *Antiquity* 35, 91–102.

GILL, J. 1987

Improved stoves in developing countries, a critique, *Energy Policy* 15 (2), 135–44.

GLEIRSCHER, P. 1996

Brandopferplätze, Depotfunde und Symbolgut im Ostalpenraum während der Spätbronze- und Früheisenzeit. In: Huth, C. (ed.), *Archäologische Forschungen zum Kultgeschehen in der jüngeren Bronzezeit und frühen Eisenzeit Alteuropas*, Regensburger Beiträge zur prähistorischen Archäologie, 2. Regensburg, Universitätsverlag (Bonn: Habelt), 429–49.

GRAHAM, J. W. 1962

The Palaces of Crete. Princeton, University Press.

HOREDT, K. – SERAPHIN, C. 1971
Die prähistorische Ansiedlung auf dem Wietenberg bei Sighişoara-Schässburg, Antiquitas, Reihe 3, Band 10. Bonn, Rudolf Habelt.

HUTCHINSON, R. W. 1962
Prehistoric Crete. Harmondsworth: Penguin Books.

HUTH, O. 1939
Der Feuerkult der Germanen, *Archiv für Religionswissenschaft* 36, 108–34.

KOSTRZEWSKI, J. 1950 (ed.)
III Sprawozdanie z prac wykopaliskowych w grodzie kultury łużyckiej w Biskupinie w powiecie żnińskim za lata 1938–1939 i 1946–1948. Poznań, Polski Towarzystwo Prehistoryczny.

KRÄMER, W. 1966
Prähistorische Brandopferplätze. In: Drack, W. – Wyss, R. (eds.), *Helvetia Antiqua. Festschrift Emil Vogt*, R. Degen, 111–22. Zürich, Schweizerisches Landesmuseum.

MARINGER, J. 1976
Fire in prehistoric Indo-European Europe, *Journal of Indo-European Studies* 4 (3), 161–86.

SCHUSTER, C. – COMŞA, A – POPA, T. 2001
The Archaeology of Fire in the Bronze Age of Romania. Bibliotheca Musei Giurgiuvensis, Monograph Series 2. Giurgiu, County Museum.

VERMEULE, E. T. 1964
Greece in the Bronze Age, Chicago and London, University of Chicago Press.

3

The Fireplace as Centre of Life

ULLA ODGAARD

Introduction

The present work offers a methodological approach to hearth studies in general and a couple of arctic hearths are shown as examples. The hearths of the Palaeo-Eskimo tradition are often well preserved which makes it possible to interpret which heating processes took place and what they meant for the indoor climate of the dwelling. The Palaeo-Eskimos made use of a versatile pyro-technology, adjustable to the most extreme conditions in areas where access to firewood is limited. An archaeological experiment in combination with a theoretical calculation on the consumption of fat shows that it was possible for the people of the Independence I culture to live through the High Arctic winter in tents with a reasonable degree of comfort. Also the symbolic aspects of hearths are discussed.

If ancient hearths could speak they might tell stories that people once were telling each other sitting around the open fire. Those stories are gone forever and the hearths are mute, but we can still learn something from them by asking the right questions. The questions could be related to fuel, heat, light and cooking, and maybe also about design and meaning in a more ideological sense.

Terminology and description of hearths

To be able to interpret the function of a hearth it is necessary that all elements are sufficiently described.

At the colloquium "Nature et Fonction des foyers préhistoriques" in Namur 1987 it was agreed upon that archaeologists should concern themselves with hearths and activities related to it equally much as they are concerned with analysing the technical and economic consequences of flint production (OLIVE – TABURIN 1989). It was further stated that even though the hearth is an evident structure, archaeological descriptions are often inexact and limited to an overall interpretation. This implicates the need for using a descriptive and precise vocabulary without ambiguities in order to establish solid basis for comparative studies (*ibid.*). It is, for example, important to be able to distinguish between the overall "hearth/fireplace" and the more precise "combustion area", since even though a feature is described in the archaeological literature as "hearth" or "fireplace" it is not always obvious where the actual combustion process has taken place.

To comply with these demands hearths should be regarded as structures composed by different elements, which can be described separately making it possible to interpret their interrelated function.

The elements a hearth can be composed of are:
- Traces of combustion: ash, charcoal, burned bone or fat;
- Moveable rocks, possibly fire-cracked; and
- Feature with area of combustion, which can be a fixed stone construction.

Traces of combustion

A combustion process that includes firewood will always leave traces of charcoal and/or ash. The probability/degree of washing out or other factors, which may have removed charcoal from a feature, should be considered already during the excavation situation.

Charcoal should be quantified. The measure can be given by volume, making it possible to estimate the amount in relation to other hearths, as for example done by SOFFER (1985) to estimate the intensity in use of different hearths.

Fire-cracked rocks

Within arctic archaeology there is a large yet still relatively unexploited potential in analysis of stone material from hearths with moveable rocks.

The documentation of fire-cracked rocks can be carried out on different levels. Firstly, a simple recording of whether fire-cracked rocks are present or not, their location at the site and in connection to which features. Secondly, weighing of the fire-cracked rocks can be done in order to provide quantification, which can be compared to other features and sites (OLSEN 1998). This kind of documentation will, however, not be precise enough to make it possible to estimate the length or intensity of the use of a site, since different types of rocks have different capacities and will be worn down at different rates when used as heating elements. This has been clearly illustrated by BUCKLEY's experiments (1990). Based on examinations of the rock type, size and degree of cracking it is possible to estimate the length/ intensity of use of an archaeological feature (BUCKLEY 1990; MARKSTRÖM 1996; ODGAARD 2001b and 2003).

To be able to tell whether the rocks collected for heat treatment were selected meticulous or picked randomly it is necessary to investigate the frequency of rock types found in the vicinity. At Head-Smashed-In Buffalo Jump in Alberta, Canada, it has been demonstrated that effort and time was spent on importing rocks, since the great majority of the fire-cracked rocks were not local (BRINK – DAWE 2003).

Typology

Contemporary research into hearths has pointed out some basic problems among others that it is only possible to compare hearths when a preliminary classification exists, including a hierarchic organisation of the descriptive traits of the features. But regarding hearths the descriptive approach may not cover the full sequence of uses and functions a hearth had in the dwelling.

What we find at the sites are hearths in a final state, where acts and activities have modified the original state. If the function of hearths remains unidentified, a classification based solely on the morphological traits can only establish formal distinctions. Consequently, a typology showing different types of hearths and not different states of use can only be established based on an understanding of the use and function of the hearths (COUDRET *et al.* 1989).

For the Arctic hearths we already use a kind of typology; for example we talk about "box-hearths" and "typical Dorset hearths" (e.g. SCHLEDERMANN – McCULLOUGH 1988). These descriptions are based on stylistic traits, which do not fully cover the function of hearths. But within demarcated geographical and cultural areas, these can be meaningful for example in defining cultural and chronological relations. Recall however that a box hearth can contain fire-cracked rocks and/or charcoal, or it can be empty reflecting different combustion processes.

Every hearth should in principle be looked upon and interpreted within its own context. I will however as a starting point suggest a simple typology, with implications regarding interpretation of the use inspired by the practical approach of PERLÈS (1977) and based on my work with hearths in general (*Table 3.1*).

Ulla Odgaard

Table 3.1. Hearth typology

HEARTHS:	WITHOUT ROCKS	WITH ROCKS	
		Fixed rocks	**Movable rocks**
PROCESS:	Open combustion Radiation	Open combustion Radiation Convection	Closed combustion Convection
RESULT:	Light and heat	Light and heat	Heat
CULINARY OPTIONS:	Broiling/grilling Boiling/cooking in pot	Broiling/grilling Roasting Boiling/cooking in pot	Roasting Boling/cooking in pot Boiling with rocks

Hearths without rocks

A hearth – where rocks are neither part of the construction nor form part of it as moveable fire-cracked elements – can, since it does not contain heating elements, only transfer radiant heat to a dwelling, and the combustion process requires good ventilation. The process will yield light, but culinary options are limited to broiling or grilling while direct boiling would be possible only if a fireproof container is available.

Hearths with fixed rocks

A combustion area situated on or within a construction of stone – beyond the radiant heat produced by the open fire – can afford convectional heat from the stones. When the combustion process has come to an end and it is possible to shut off the ventilation of the dwelling by closing the smoke-hole and/or the entrance. The first part of the process will provide for light while the second part will leave the dwelling in darkness unless other sources of light are in use. In addition to broiling, grilling and boiling in a pot, in hearths with fixed rocks it is also possible to roast by placing the food directly on the hot rocks.

Hearths with moveable rocks

Hearths with a content of moveable rocks should be interpreted after considering the relation of the rocks to charcoal. Rocks that are not in context with charcoal could have been transported from the heating source into a dwelling where all air-channels are tightly closed. The process of heat transmission would be convection, affording an even temperature in the room. Fire-cracked rocks that are clean and found outside the context of soot and charcoal could moreover have been used for boiling of liquid in a not necessarily fireproof container. Rocks that are mixed with charcoal were probably heated on the spot in an open fire. When the fire died out the rocks continued to function as heating elements, affording moderate heat for a longer time than the embers. During the process it would have been possible to boil/heat liquid in a fireproof container put directly on or among the rocks, and flat rocks could have been used as frying pans.

Further analyses

Experimentation plays an important part in the ongoing research on hearths, as does studies of ethnographic documented hearths. Experiments can control the efficiency of different hearth-arrangements and can throw light on both given conditions and deliberate choices. On this background experimental models can be developed, which in connection with the archaeological context make it possible to distinguish between conditions and choices.

More sophisticated technical analyses, as shown by primarily the French archaeologists in "Nature et fonction des foyers préhistoriques" (OLIVE – TABURIN 1989), should be applied to the study of the hearth. It is an obvious area for archaeometric methods, and technical analysis of the construction and content of the hearths can form part in the analysis as information, which can support the interpretation of the practical aspects. Modern research includes lipid-analyses, micro morphology, charcoal-analyses and other analyses. But without archaeological interpretation, which includes analogy, research provides just more accurate descriptions of hearths.

We need to synthesize, and I suggest that the methods and considerations presented in the present work can be made use of in this connection. To illustrate this point I will present interpretations of a couple of Palaeo-Eskimo hearths.

Who were the Palaeo-Eskimos?

Extreme cold, ice and scarcity of firewood! These were the options for the first people to enter the arctic areas of Eastern Canada and Greenland around 2500 BC, after the last ice age finally had let go of these areas. We call the people the Palaeo-Eskimos even though they are not directly related to the Thule Eskimos, who entered the same area around 1000 AD. Their place of departure was most probably in Central Siberia, from where they spread rapidly, literally without leaving traces on their way to high arctic Canada and Greenland. They were highly mobile and able to follow the flocks of musk ox and caribou but they also hunted birds, fish and seal. They brought with them the so-called "arctic small tool tradition", which links them to early Neolithic cultures of central Siberia. Also bow and arrow was introduced to the New World by the Palaeo-Eskimos (MCGHEE 1996).

Indoor climate

When it is possible to deduce from the archaeological remains which kind of superstructure a dwelling had, it is also possible to make a theoretical reconstruction of the indoor climate of the dwelling.

The preferred dwelling of the Palaeo-Eskimos was the tent, and what is left of them is actually limited to the floor and outline. The outline is either a tent ring (ring of rocks) or a low wall of either gravel or turf.

A tent is never just a bundle of poles covered with a skin. Tents are also designed. Tents of the historic nomadic people in the Arctic- and Sub-Arctic regions were noted for their strength and stability. The demanding conditions in the Arctic suggest that prehistoric tents were just as refined and specialized as tents known from ethnography. Ethnographically known tent types of the Northern Hemisphere are quiet few but widely distributed, and all testify to be of great age (FAEGRE 1979).

Strong similarities can be seen between the floor plan of prehistoric and ethnographically known tents from the Arctic and Sub arctic areas. This suggests that the architectural design of the prehistoric tent dwellings were much alike or even identical with types of tents still in use among nomadic peoples in Arctic and Sub Arctic during the 20th century (ODGAARD 1995).

In the following will be given a couple of examples of interpretation of Palaeo-Eskimo dwellings with:

- A box-hearth from the most extreme situation.
- A summer hearth.
- Two hearths.

A box hearth from the most extreme situation

The earliest Palaeo-Eskimos in the High Arctic belonged to the Independence I culture (2500–1900 BC). The only source of heat in their tents came from the hearth, since no oil/blubber-lamps have yet been found in this early phase. Consequently the inhabitants must have paid a great deal of attention to the functioning of the hearth.

Some of the hearths are box-hearths, built of slabs on edge and the size is most often around 40 x 40 cm, some hearths are in a mid-passage (or axial feature), and within a tent ring (*Figures 3.1a* and *3.1b*). The contents of the box-hearths can vary, but I will concentrate on those suggested by SCHLEDERMANN – MCCULLOUGH (1988) to be traces from the colder time of year. "Some of the more well-preserved hearths contain an interior layering of irregular-sized flat stones mixed with charred bone, wood and grease-saturated sand" (*ibid.* 6). They were maybe from the most extreme situation during the Arctic winter of two months total darkness and temperatures below –30° C.

How was it possible to survive during high Arctic winter in a tent with a fireplace without either dying from too much smoke or from the extreme cold that sufficient draft would allow in? SCHLEDERMANN – MCCULLOUGH (1988, 20) suggest that the tents were abandoned by mid-winter in favor of temporary snow house camps on the sea ice or snow-banked tents on land. But the absence of lamps does not support this hypothesis. KNUTH (1967), MCGHEE (1979, 1996) and MAXWELL (1985) suggest that winter was spent asleep in a kind of torpor in a frozen tent. MCGHEE (1996, 64) writes: "... it is difficult to imagine how their winter lives could have been very different from this portrayal. The hard-won fuel resources that they did accumulate must have provided only for small and occasional fires, enough to thaw food and melt ice for water but not to provide heat for human warmth." McGhee considers firewood and musk ox bones to be the only fuel sources (MCGHEE 1996, 50), but according to SCHLEDERMANN (1990, 50) the site locations in the Eastern High Arctic points to preference for open water sea mammal hunting, particularly seals. Seal bones are found at Independence I sites both at Ellesmere Island (*ibid.*) and in High Arctic Greenland (e.g. GRØNNOW – JENSEN 2003).

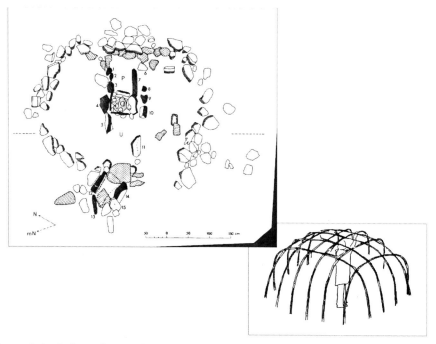

*Figure 3.1a. Independence I dwelling from Gammel Strand Nord, Jørgen Brønlund Fjord,
High Arctic Greenland (from GRØNNOW – JENSEN 2003: Figure 5.64) (KNUTH 1967).
Illustration of suggested reconstruction from FAEGRE (1979: 134).*

The suggestion of the over wintering of the earliest Palaeo-Eskimos in the cold
tent with only occasional heat from the hearth is often presented. It was however
not obvious how the above described hearth-arrangement could have functioned
and affected the situation in the tent, and since there was no ethnographic analogy
to this type of hearth I carried out a contextual archaeological experiment. There
are two different experimental approaches in archaeology: the "controlled"
archaeological experiment and the "contextual". The controlled archaeological
experiment is related to the way experiments are made in natural sciences with a
basic rule of changing one parameter at the time and keeping the others constant.
The contextual archaeological experiment does not claim to control the variables,
but searches for identical situations to be inspired from (RASMUSSEN 2001, 6).
"Its main contribution is the supplying of an interpretative framework, rather than
providing proofs. Having gained new insights, the experimenter will be able to go
back to the archaeological record" (*ibid.* 8).

An experimental hearth was built following Schledermann and McCullough's
description (fist size rocks on a layer of gravel/sand inside a box of slabs on edge).

*Figure 3.1b. Detail of mid-passage with box hearth from fig. 1a
(Ill. GRØNNOW – JENSEN 2003: Figure 5.65).*

The idea from the beginning was that it had to be a very controlled combustion process since the tent was quite small, and that the smoke from the fire should be kept to a minimum. This process was because of the possibility of making the smoke hole smaller, and keeping the heat inside the tent.

The experimental hearth was heated with firewood and seal bone (which burned very well together with wood). In order to understand how and why the gravel inside the archaeological hearths got saturated with fat I experimented with various methods of melting fat in the hearth. They were all very smoky until I tried to place the fat with a wick of moss (as in a lamp) in the hearth. It functioned quite well when the hearth was warm, but not if it was hot, because then the fat started to fry and send out black smoke. I could not make it function in a cold hearth either.

Fifty grams of fat burned for 25 minutes making very little smoke (*Figure 3.2*). A possible explanation for the gravel/sand in the archaeological hearths is that it is convenient to have a sand reservoir for the melted fat, if you have to stay for a period of some length close to a hearth where fat is part of the fuel. In a lamp without drainage the wick will get swamped unless the excess of melted fat is removed by tipping the lamp (BAUNE 1987).

Following this new theory on the Independence I winter box-hearth having a kind of lamp-function, it is possible to make a theoretical calculation on the consumption of fat by means of technical standards (*Table 3.2*). The calculations are proposed and made by Ulrik Henriksen, Associate research professor at the Technical University of Denmark (the full calculations in ODGAARD 2001b).

Figure 3.2. Experimental hearth with the suggested arrangement in function.

Theoretically the standard situation in an Independence I winter tent (inspired by *Figure 3.1*, from Gammel Strand, Jørgen Brønlund Fjord, Greenland) might have been as follows:

- Tent = dome shaped, diameter 4 m, max. 2 m high
- Tent cover = 2 layers of caribou or musk-ox skin
- Tent floor = 2 layers of caribou or musk-ox skin
- Number of persons = 6
- Outdoor temperature = -30° C
- Indoor temperature = +8° C, which is a temperature reported ethno-graphically to be pleasant, if the inhabitants are wearing skin clothes, and it is also warm enough for ice to be melted for drinking water.

Table 3.2. Energy calculations (by Ulrik Henriksen).

ENERGY OUT	Situation 1	Situation 2
Tent cover	669	669
Tent floor	70	70
Ventilation	1186	165
Air for combustion	15	15
Melting of ice	46	46
Energy out – total	1986	965

ENERGY IN	
Heat from people	648
Burning of fat (120 g pr. hour)	1250
Energy in – total	1898

Situation 1 = Loss of heat at "normal" ventilation – Watt
Situation 2 = Loss of heat at a minimal ventilation – Watt
The numbers refer to energy in Watt = j/second.

Since energy loss through ventilation is of considerable importance two calculations with different ventilation factors are made (see *Table 3.2*).
- Situation 1. The lowest renewal of the air necessary for people feeling comfortable.
- Situation 2. Least necessary amount of air for survival (as in submarines, spaceships and the like).

Heat is lost through the tent cover and floor, necessary ventilation, air for combustion and melting of ice for two litres of drinking water per person. In the first column (*Table 3.2*) the energy loss is calculated at a situation with comfortable ventilation, while in the second column it is with minimal ventilation like the situation in submarines. Heat is gained through body heat and through the combustion of fat in the hearth.

In situation 1 (being the most probable) the energy gained is close to the energy lost. This situation would result in a fat consumption of 120 g per hour, 2.88 kg per day, or for the two dark months, 175 kg altogether.

In situation 2 less fat is required – only 740 g per day, or 45 kg for the two dark months. If the box-hearth had a continuous function in the way I have suggested, the Independence I people did not have to experience severe frost in the tent during winter, since 175 kg of seal fat (around 12 seals) would be an easy task for a skilled hunter, and when stored amounted to around only 0.175 cubic metres.

For this experiment and calculations seal fat was chosen because of its qualities as fuel (BAUNE 1987), and because bones from seal have been found at Independence I sites (SCHLEDERMANN 1990), but fat from other animals e.g. musk ox could be used as well. Physico-chemical analyses of residue samples from the archaeological hearths could throw light on this question.

An argument against the Independence I people to have over wintered in tents in the High Arctic is the small size and low number of meat caches (SCHLEDERMANN – MCCULLOUGH 1988, 20). But even though the Independence I people are often described as highly mobile, they were not necessarily opportunists relying exclusively on fresh meat. From ethnography many examples are known of meat being dried for later use – either by air or sun or by drying over fire or smoke (e.g. DRIVER – MASSEY 1957, 243).

Both KNUTH (1967) and PLUMET (1989) point out the possibility that the stone built mid-passages may have functioned as meat caches. The meat could have been preserved in the same way as the Plains Indian's "jerky" (meat cut in very thin slices). It has the advantages of taking up less space than fresh meat, it is more light weight and can be kept longer and yet still has a high nutritional value (LAUBIN – GLADYS 1989, 149).

Among others, the Coastal Salish Indians lived during the winter on dried food that was boiled for soup or stew by the use of hot rocks. It is however unlikely that the box-hearths, with their content of fat-saturated gravel and sooty rocks, were used for heating rocks for boiling of soup and cooking pots were not used by the Independence I culture.

SCHLEDERMANN – MCCULLOGH mention the possibility of roasting in the hearths by placing the meat directly on the hot rocks (1988, 6). Some times the Plains Indians ate the dried jerky as it was, but they preferred it cooked. One of the cooking methods was to roast it until the fat began to show and the meat became a brown colour like broiled fresh meat. According to LAUBIN – GLADYS (1989, 153) eating a peace of jerky prepared this way is a very crunchy, but tasty experience.

A summer hearth

An example of a summer hearth is also from the Independence I culture and from Lakeview Site on Ellesmere Island in the Canadian High Arctic (SCHLEDERMANN 1990, *Figure 3.3*). The hearth is composed of small, round stones put directly on the exposed bedrock-surface. The tent ring (of gravel) is oval 4.5 x 3.5 m, suggesting a Sami kåta tent type or a so-called "purlin tent" with a frame of two cones connected by a purlin pole lengthways, which would make enough room for an open fire. The fire would provide sufficient light for different kinds of handicrafts, which is reflected by the finding of microblades, burin spalls, flake knifes and points within the tent ring. Now and again an open fire on a flat surface requires cleaning out. In this instance charcoal and ashes were occasionally scattered outside the fireplace itself.

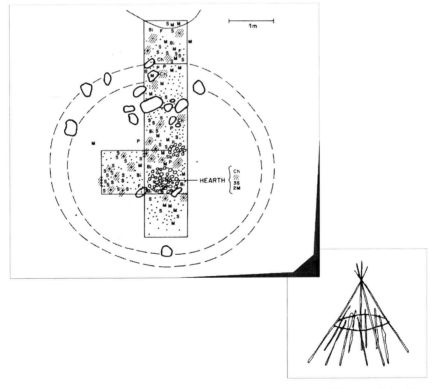

Figure 3.3. Independence I dwelling from Lakeview Site, Ellesmere Island (from SCHLEDERMANN 1990, Figure 10). Illustration of suggested reconstruction (from FAEGRE 1979, 132).

The combustion area of this fireplace was placed directly on the rock – and the combustion process was an open fire heating the tent by radiation. Most of the heat vanished together with the smoke up the smoke hole, suggesting this dwelling to be from a warmer period of the year.

The cooking options were broiling over flames, grilling over embers and boiling. The small round and flat stones in the hearth and in a small pile nearby indicate that cook-stone-technology was applied for boiling as suggested by SCHLEDERMANN (1990).

Rocks were used as heating elements through the Palaeo-Eskimo tradition, but the method and intensity vary. Cook-stone technology in culinary practises is known from different ethnographic sources, for example the Coastal Salish Indians, who cooked with hot rocks during the winter when stored dried food was prepared. Round volcanic rocks were heated in the coal of the fire and put in water in a basket or a box of cedar, where the rocks would bring the water to the boiling point. Their fireplaces for heating rocks are described as "small and simple" (BATDORF 1990), and it was necessary to use hard wood for creating the intense heat, which only glowing embers produce. The Coastal Salish Indians placed the rocks close to the fire and on top of other large and flat rocks in order to keep the boiling stones as free from the ashes as possible. The rocks were constantly rotated and moved around for steady heating, thrown into the basket and removed again to be heated in the fire after cooling. Before putting the rocks into the basket they were quickly dipped in another basket with clean water without reducing their heat considerably (*ibid.*).

This method of using rocks for boiling fits well with the round hearth from Lakeview Site with the small and round rocks. The rocks being round suggest that they were not fractured by repeated heatings, and that they were probably selected for their present size. Their smallness indicates a heating of liquid, as smaller rocks are effective for this exact purpose because of the quick absorption and liberation of heat. Large rocks absorb and liberate heat slower, which make them more suitable for other culinary purposes and for room heating (MARKSTRÖM 1996). There are however no traces of cooking pots from the Independence I culture. The small rocks suggest instead that organic containers, which were not preserved such as those of skin or weaved plant material, were used.

A closer study of the small rocks, interpreted as "boiling rocks", could identify the type of rocks, which would make us able to estimate the intensity/length of use of the hearth.

Two hearths

The tradition of having two hearths in the dwelling reaches far back in the Palaeo-Eskimo tradition, where one central and one non-central hearth (from late Saqqaq and onward this can be a lamp) often occur. But it is not until late Dorset (in Greenland ca. 750–1300 AD) that we find the second fireplace incorporated in the mid-passage, as for example at the Polaris site, feature 1 in Northern Greenland (*Figure 3.4*). The rounded-rectangular shape of the ground plan, the relatively narrow benches (around 125 cm) at the sides, and the location of the hearths in the broad mid-passage make it likely that the tent was dome-shaped. A lamp-support was found at one end of the mid-passage and at the other end a hearth for an open combustion. The lamp-support is a heavy rock of 30 kg with an even surface and an oval depression rimmed with black-crusted blubber. Also along the sides of the rock the soil was black, probably from overflowing blubber. This could be noted to a depth of 15–20 cm in the soil (GRØNNOW 1999, 51).

The lamp to have stood at the lamp-support represents a prolonged combustion process, which by radiation could provide heat and light, while clothes or other things could be dried above it. If the smoke hole and tent doors were closed this arrangement would provide the tent with a warmer temperature not only close to the lamp but also in the whole tent due to convection. When a warm object is placed in a stagnant fluid – as cold air – the reduction of the density in the heated fluid close to the warm object will result in movement. The heated fluid will move upwards and be replaced by colder air, which again will be heated – making a flow of warm air. In other words, heating by convection – a process without smoke (STEWART 1987).

If the moveable andirons found close to the structure were placed on each side of the lamp-support they could have supported a cooking-pot – heated by the lamp. The other hearth represents an open combustion facility used for a shorter process in connection with culinary practises with the smoke hole open. Here the moveable andirons could have functioned as support for spits used in broiling over flames or grilling over embers, or without the andirons for roasting on a flat rock (*Figure 3.5*).

The two hearths in combination provided the dwelling with a source of light, a stove and a multi-kitchen supplying all the culinary needs a Palaeo-Eskimo gourmet could ask for.

When most of the structures discussed above were carefully excavated, focus was not on the pyro-technology, and details about the type and size of stones, and clear evidence of combustion was not always recorded. Attention to this kind of evidence could provide more reliable results than I have been able to set out in the above.

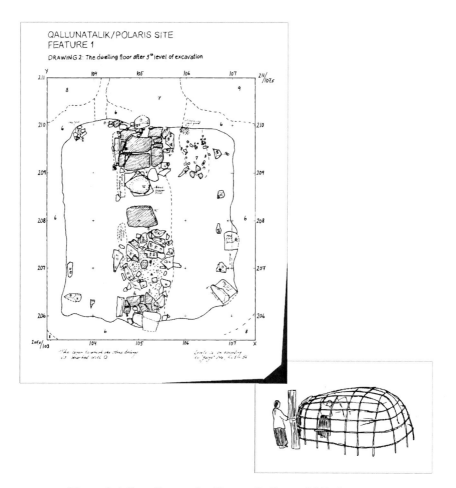

Figure 3.4. Late Dorset dwelling at Qallunatalik/Polaris site,
High Arctic Greenland (GRØNNOW 1999: Figure 44).
Illustration of suggested reconstruction (from FAEGRE 1979: 140).

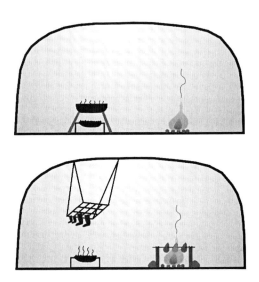

Figure 3.5. Reconstruction of hearth arrangements in late Dorset dwelling Qallunatalaik/Polaris site.

Introduction of fireproof pots

The most important innovation in connection to the pyro-technology of the Palaeo-Eskimos seem to have been the introduction of fireproof cooking vessels of soapstone, which is also reflected in the amount of fire-cracked rocks at the sites.

In the above examples from the Independence I culture in the High Arctic fire-cracked rocks were present at a limited scale, but further south in western Greenland the contemporary Saqqaq culture (c. 2400–800 BC) used hearths with up to several hundred fire-cracked rocks. One was documented in a way that made it possible to interpret the function and length/intensity of use (ODGAARD 2001b and 2003). During the Saqqaq culture lots of driftwood was available on the beaches in Western Greenland (GRØNNOW 1996) enabling the Palaeo-Eskimos to apply stone heating technology at a greater scale – a technology that requires lots of fuel. Fire-cracked rocks found in – or in connection to – fixed hearth-structures are common in this period. According to JENSEN (1998) box-shaped and circular hearths with fire-cracked rocks is exclusively a Saqqaq-phenomenon, while Dorset heating and cooking activities were closely related to the use of soapstone vessels (JENSEN 1998, 76). Yet according to MØBJERG

(1999) vessels and lamps of soapstone were already introduced during Saqqaq times, which can be divided into an older and a younger phase. In the Sisimiut area there are many box-hearths with fire-cracked rocks, but no evidence of vessels or other objects of soapstone from the older phase. However, from the younger phase large numbers of fragments and objects of soapstone including vessels and lamps are present (MØBJERG 1999, 462). Møbjerg suggests that a decrease in availability of driftwood triggered a change to an increased emphasis on blubber for fuel (*ibid.*). This could be part of the explanation since the large amount of driftwood that had piled up on the beaches prior to the arrival of the Saqqaq-people was not an inexhaustible source (GRØNNOW 1996). But the same pattern can be seen in Northern Scandinavia, where thousands of sites with fire-cracked rocks have been reported and where intensive use of fire cracked rock technology continued through a long period (from around 5500 BC to 3300 BC). The technique was still used though not to the same extent around 1900 BC but around 1300 BC (same date as in Greenland) the pattern changed to only few fire-cracked rocks and well-constructed hearths, which can be associated with the introduction of ceramics (BROADBENT 1979).

Both in Greenland and in Northern Scandinavia the application of fireproof cooking vessels meant that a more efficient type of heat transmission process could be employed. A combustion process involving heating of fire-cracked rocks requires much labour and it is unavoidable that energy is lost during transmission of the heat from the fire and the embers through the rocks to the food. If instead the cooking vessel is placed directly above or in the fireplace both labour and energy is saved.

The Hearth as a Symbol

In the above I have solely touched upon the practical aspects of the hearths, but we should also consider what hearths meant to the Palaeo-Eskimos. I have earlier suggested the Palaeo-Eskimo mid-passage to be symbols of the shamanistic clan river, known from shamanistic seances in Siberia (ODGAARD 1998 and 2001a; MØBJERG – ODGAARD 1999). In a survey of religious ideas in connection to hearths from different cultures from Siberia, Canada, Northern Scandinavia and Europe I have found, that trans-culturally the fireplace is regarded as a gate to other worlds. During this opening offerings can be given to the dead or to the gods, and rites of feeding the fire or the spirits are known from many cultures (ODGAARD 2001b).

The god of the hearth is often a woman who guards the family and the clan, and who assists at childbirth. The hearth goddess is a widely spread phenomenon known in Europe as the Greek "Hestia" or the roman "Vesta" (e.g. JONES – PENNICK 1995). In South Scandinavia she was called "Eldborg" and in northern Scandinavia "Sarakka" (BÄCKMAN 1984). In Siberia she also had different names and could have the shape of an old woman or a small girl. But everywhere she was the mother and protector of the clan (e.g. SIMCENKO 1978).

Another widespread idea is that ancestors or other spirits, who can help to success during hunt, can be contacted through offerings in the fireplace, and that these spirits have their seat close to the fire (e.g. TUGOLUKOV 1978 and TANNER 1985).

Among the historical Siberian Tuvans the helping spirits could be fed with special spoons (VAJNŠTEJN 1978), and small spoon-formed devices were also among the Palaeo-Eskimo's possessions. The illustrated specimen (*Figure 3.6*) found close to a Dorset hearth on Devon Island (MARGARETH BERTULLI, *pers comm.*) might be such an implement.

Figure 3.6. "Spoon-formed device" excavated close to Dorset hearth on Devon Island.
Photograph: Margareth Bertulli.

Most numerous among the archaeological features in the Arctic landscape are hearths and traces of dwellings, which still today in the High Arctic nearly look like monuments and make it a "socialized" landscape that preserves history and meaning (TILLEY 1994). The Palaeo-Eskimos made the Arctic their home, and the hearths and the mid-passages must have been important symbols and reminders of the pyro-technology, culinary practices, clan-connections and spiritual power, whether it was one built by themselves the previous year or by their ancestors thousands of years ago.

Bibliography

BATDORF, C. 1990
Northwest Native Harvest, Canada.

BAUNE, S. A. De 1987
Palaeolithic lamps and their specialization: a hypothesis. *Current Anthropology,* Vol. 28, No. 4, 569–577.

BRINK, J. W. – DAWE, B. 2003
Hot Rocks as Scarce Resources: The Use, Re-Use and Abandonment of Heating Stones at Head-Smashed-In Buffalo Jump. *Plains Anthropologist* Vol. 48, Number 186, 85–104.

BROADBENT, N. 1979
Coastal Resources and Settlement Stability. A Critical Study of a Mesolithic Site Complex in Northern Sweden. Uppsala, Aun 3.

BUCKLEY, V. M. 1990
Experiments using a reconstructed fulacht with a variety of rock types: implications for the petromorphology of fulachta fiadh. In: Buckley, V. (ed.), *Burnt Offerings. International Contributions to Burnt Mound Archaeology,* 170–172.

BÄCKMAN, L. 1984
The Akkas. A study of four goddesses in the religion of the Saamis (Lapps). In: Tyloch, W. (ed.), *Polish Society for the Science of Religions. Current progress in the methodology of the science of religions.* Warsaw.

COUDRET, P. – LARRIERE, M. – VALENTIN, B. 1989
Comparer des foyers : une entreprise difficile. In: Olive, M. – Taborin, Y. (eds.), *Nature et fonction des foyers préhistoriques. Mémoires du Musée de Préhistoire d'Ile de France* n° 2, 37–46.

DRIVER, H. E. – MASSEY, W. C. 1957
Comparative studies of North American Indians. Transactions of the American philosophical Society, held at Philedelphia for promoting useful Knowledge. New Series; Vol. 47, Part 2. Philadelphia.

FAEGRE, T. 1979
Tents: architecture of the nomads. London, Murray.

GRØNNOW, B. 1996

Driftwood and Saqqaq Culture Woodworking in West Greenland. In: *Cultural and Social Research in Greenland 95/96. Essays in Honour of Robert Petersen.* Nuuk, Ilisimatusarfik/Atuakkiorfik, 73–89.

GRØNNOW, B. 1999

Qalunatalik/Polaris site. In: Appelt, M. – Gulløv, H. C. (eds.), *Late Dorset in high arctic Greenland : final report on the Gateway to Greenland project.* Copenhagen, Danish Polar Center Publication No. 7. Danish Polar Center, 42–62.

GRØNNOW, B. – JENSEN, J. F. 2003

The Northernmost Ruins of the Globe. Eigil Knuth's Archaeological Investigations in Peary Land and Adjacent Areas of High Arctic Greenland. Meddelelser om Grønland. Man & Society 29.

JENSEN, J. F. 1998

Dorset dwellings in West Greenland. *Acta Borealia,* Volume 15, 59–80, Oslo.

JONES, P. – PENNICK, N. 1995

A history of Pagan Europe. London.

KNUTH, E. 1967

The Ruins of the Musk Ox Way. *Folk* 8–9, 191–219.

LAUBIN, R. – GLADYS, H. 1989

The Indian Tipi. Its history, construction and use. Norman.

MARKSTRÖM, M. 1996

Skärvsten – vad är det? En experimentell studie. CD – uppsats i Arkeologi, Vt 1996. Umeå Universitet, Institutionen för arkeologi.

MAXWELL, M. S. 1985

Prehistory of Eastern Arctic. New York.

McGHEE, R. 1979

The Palaeoeskimo occupations at Port Refuge, High Arctic Canada. Archaeological survey of Canada, paper no. 92. Ottawa.

McGHEE, R. 1996

Ancient People of the Arctic. Canadian Museum of Civilization.

MØBJERG, T. 1999
New adaptations strategies for the Saqqaq culture, West Greenland. *World Archaeology* 30 (3), 452—465.

MØBJERG, T. – ODGAARD, U. 1999
Ild og ånd i en verden af sne og is. Hvem var palæoeskimoerne? *Menneskelivets mangfoldighed. Arkæologisk og antropologisk forskning på Moesgård*, 335–342.

ODGAARD, U. 1995
Telte i arktiske miljøer. Rekonstruktioner og ideologi. Cand.phil.speciale, Institut for Arkæologi og Etnologi, København.

ODGAARD, U. 1998
The Arctic Midpassage and its Religious aspects. In: Anderson, A. m.fl. (ed.), *The Kaleidoscopic Past.* Gotarc Serie C, No. 16, Gøteborg, 462–68.

ODGAARD, U. 2001a
Palaeo-Eskimoic Shamanism. In: Vestergaard, T. A. (ed.), *North Atlantic Studies. Shamanism and Traditional Beliefs*. Vol. 4, no. 1 + 2, Aarhus, 25–30.

ODGAARD, U. 2001b
Ildstedet som livscentrum. Aspekter af arktiske ildsteders funktion og ideologi. Ph.d. afhandling, Forhistorisk Arkæologi, Moesgård, Århus.

ODGAARD, U. 2003
Hearth and Home of the Palaeo-Eskimo. *Études/Inuit/Studies. Palaeo-Eskimo Architecture.* vol. 27 (1–2).

OLIVE, M. – TABURIN, Y. 1989 (eds.)
Nature et fonction des foyers préhistoriques. Actes du Colloque International de Nemours 1987. Mémoires du Musée de Préhistoire d'Ile de France, nº 2.

OLSEN, B. 1998
Saqqaq housing and settlement in southern Disko Bay, West Greenland. *Acta Borealia,* Volume 15, Oslo, 81–128.

PERLÈS, C. 1977
Préhistoire du feu. Masson.

PLUMET, P. 1989
Le foyer dans l'Arctique. In: Olive, M. – Taborin, Y. (eds.), *Nature et fonction des foyers préhistoriques. Mémoires du Musée de Préhistoire d'Ile de France,* n° 2, 313–326.

RASMUSSEN, M. 2001
Experiments in archaeology – A view from Lejre, an "old" experimental centre. *Zeitschrift für Archaeologie und Kunstgeschichte,* Schweiz, 3–10.

SCHLEDERMANN, P. 1990
Crossroads to Greenland. 3000 Years of Prehistory in the Eastern High Arctic. The Arctic Institute of North America.

SCHLEDERMANN, P. – McCULLOUGH, K. 1988
Hearth of Darkness: Structural variability in ASTt dwellings in the Canadian High Arctic. Paper Presented at the 21st Chacmool Conference, November 10–13.

SIMCENKO, Ju.B. 1978
Mother Cult among the North-Eurasian Peoples. In: Diószegi, V. – Hoppál, M. (eds.), *Shamanism in Siberia.* Budapest, 503–514.

SOFFER, O. 1985
The Upper Palaeolithic of the Central Russian Plain. Orlando, Academic Press.

STEWART, B. (and others) 1987
Improved Wood, Waste and Charcoal burning Stoves. A practitioners' manual. London, IT Publications.

TANNER, A. 1985
Bringing Home Animals. Religious Ideology and Mode of Production of the Mistassini Cree Hunters. Social and Economic Studies, no. 23. Memorial University of Newfoundland.

TILLEY, C. 1994
A phenomenology of landscape : places, paths and monuments. Explorations in anthropology. Oxford.

TUGOLUKOV, V. A. 1978

Some Aspects of the Beliefs of the Tungus (Evenki and Evens). In: Diószegi, V. – Hoppál, M. (eds.), *Shamanism in Siberia.* Budapest, 419–428.

VAJNŠTEJN, S. I. 1978

The erens in Tuva Shamanism. In: Diószegi, V. – Hoppál, M. (eds.), *Shamanism in Siberia.* Budapest, 457–468.

4

The Symbolic Meanings of Cremation Burial

PAULA PURHONEN

The study of burial customs offer an important basis for investigating the beliefs concerning the afterlife held by societies which left no written records. However, these customs also reflect the culture of death of contemporary societies. The form of burial can be defined as a practice based on a community's beliefs, rules and traditions, carried out in order to deal with the dead body and lead the deceased to the netherworld according to the ritual customs considered appropriate by that community. Of the burial forms it is inhumation and cremation, often involving structures outside and inside the grave that have left their mark in the landscape and/or the soil. We do not know, however, what proportion of the deceased in different prehistoric periods has actually received a ritual burial, because all the burials cannot be traced and it is often impossible to date the unfurnished graves.

Although the ritual of burial essentially concerns the individual, a sense of community is still strong in the attitudes towards death, in the patterns of notification of the death and especially in the actual burial practice. In pre-Christian world view the ritual of burial was a way of confirming the crossing of frontier between the world of the living and the afterworld, but even after the adoption of Christianity people followed various rituals in order to influence the deceased's situation in the netherworld (PURHONEN 1998, 159–169). The burial ritual still carries an important meaning in terms of the community, functioning as an opportunity to evaluate the deceased and thus the greatness of the loss suffered by the community, in addition to being a manifestation of the status of the family to the wider community. On the other hand, a burial without the appropriate rituals, or in a place other than the separated sacred place, the cemetery, has often signified the eviction of the deceased from the community, while the desecration of a grave, for example, is still considered to be an insult to the whole community that the deceased in their life belonged to. Therefore, the places and customs connected with death are loaded with meanings – conscious and unconscious. One can attempt to trace and interpret these meanings by investigating the features of individual graves and cemeteries. However, since the culture of death with its underlying mental models and ritual practices based on these beliefs are the structures of culture that are the slowest to change, the traditional meanings of individual burial customs may have been forgotten or have lost some features, or perhaps have been wholly replaced by others. It appears impossible to answer the question of how familiar the individual members of community have been with the original meanings of the various features of their community's tradition in the course of history.

In our culture, the primary function of cremation burial is to reduce the body of the deceased in order that it can be interred in the smallest possible space. It is clearly more practical than inhumation, nevertheless the latter practice remains the most common form of burial. In the prehistoric period, cremation and inhumation were not the only forms of burial. Open-air burial and water burial were probably also known, but for obvious reasons these have left no archaeological traces. We tend to explain our own choice of burial form in terms of practicalities and rarely analyse the death-related beliefs, ideas, intentions, and values that underlie our own behaviour. However, we regularly associate prehistoric burials with a wide variety of symbolic meanings and levels of meaning as a matter of course. Burial customs are presumed to reflect, among other things, religious and cosmological beliefs concerning the afterlife, what happens to people after they die, what the World of the Dead is like and where it is located, how one gets there and who is admitted, how the living can influence the status of the Dead in the netherworld, what kind of relationship exists between the living and dead members of the community, and how the latter can affect the well-being of the former.

ANNA-LEENA SIIKALA (1992) has suggested that ideas of the afterlife can be divided into two groups, depending on whether the Dead are thought to dwell in a cemetery or in a distant Land of the Dead. Though this division cannot be applied as such to the Finnish archaeological record, it can be utilised as a starting point for the study of the patterns of thought relating to the cult of the Dead and the other cosmological questions referred to above. The so-called distant-netherworld pattern could be considered especially typical of arctic hunter-gatherer cultures. It reflects a fearful avoidance of the spirits of the Dead and is well suited to serve as an ideological background for pit burial and open-air burial. Other customs related to this pattern of thought could be the use of red ochre and fire in the burial ceremony, and also the practice of covering the corpse with stones. Fear of the Dead would also have led to the avoidance of burying corpses in campsites and caused people to prefer places such as islands, which were separated from the living by running water. The proximate-netherworld pattern, on the other hand, may be considered typical of farming cultures, where the relationship with the Dead was usually a positive one as ancestors were seen as protectors of the lineage's prosperity and guarantors of fertility. Therefore, the Dead were preferably kept close to settlements, which by the Iron Age had already begun to develop into permanent villages (SIIKALA 1992, 108–117).

In Finland the earliest signs of fire being connected with the burial ritual date from the Stone Age. Both Mesolithic and Neolithic red ochre graves have produced evidence of fire in the form of small, temporary hearths. The only actual Stone Age cremation burial known to date was unearthed at the Comb Ceramic site of Vaateranta in Taipalsaari, in south-eastern Finland. Though the burial contained no less than 1400 grams of burnt bone (RÄTY 1995, 165, 167), the find context is so unclear and the recording so inadequate that the find cannot be dated reliably.

Therefore, the oldest unequivocal cremations only date to the late Bronze Age, even though the decisive change in views concerning death had apparently already taken place at the beginning of the Bronze Age. At this time, the coastal population had at least partly given up the old tradition of pit burial and adopted the Scandinavian tradition of building burial cairns. The latter consisted of interring the Dead on high cliffs close to water and covering the graves with stone cairns, some of which were very large. Another change took place during the III/IV Period of the Scandinavian Bronze Age, when inhumation was replaced by cremation, the size of the cairns decreased, and they began to be built on soil rather than rock foundations, close to arable land. This phenomenon has been thought to reflect changes not only in the cult of the Dead and other religious views but also changes in the economy and the social structure (SALO 1984, 139–140; EDGREN 1992, 119).

The cairn-building tradition itself has been thought to reflect the importance of lineage rather than the individual. The cairns have been seen as monuments to the kin group and manifestations of its existence, its might, and its fame (SALO 1981, 125). Since grave goods are nevertheless rare in Bronze Age graves and since settlements relating to cairns have only been identified in a few exceptional cases, little data has been available for the study of the religious meanings and ideas that underlie the burial customs. However, it may be possible to study this burial pattern by looking at Iron Age cairns containing cremation burials, since these have produced a richer selection of archaeological materials. The practice of cremating the dead before burial continued in the area of the farming culture until the eleventh century AD, with the exception of a small region in Western

Finland (Eura-Köyliö-Yläne). Here, inhumation burial was already adopted in the sixth century AD. Cremation was finally given up only when Christianity established its position, which occurred during the thirteenth century in Western Finland (PURHONEN 1998, 137).

I have, in a previous connection, dwelt upon the symbolic meanings of Iron Age cremation burials as represented by the Merovingian Period cairn cemetery of Laitila Vidilä Vainionmäki in south-western Finland (PURHONEN 1996, 119–129). This cemetery provides excellent possibilities for studying Iron Age ideas of the netherworld, and perhaps even a larger part of the cosmology, on the basis of structural features of the site. In the following, I wish to present certain features that appear in the Vainionmäki material and to offer some interpretations of their cultural, religious, and mental background. Towards the end of the paper, I shall ask whether the burial customs of our own time have really become devoid of all of the kinds of meanings that we are prepared to ascribe to the ancient burials, and if so, has anything taken their place?

The cemetery at Vainionmäki first came to light in 1985 when a hillock or mound of moraine situated among fields was being levelled and the removed earth was found to contain "weapons of old appearance". Archaeological excavations lasting 10 years were initiated, being concluded in 1994 (SAUKKONEN 1996, 19–29). The research results have been published in 1996 by a team of six archaeologists, a numismatist, an osteologist and a botanist (VAINIONMÄKI 1996).

The remaining section of the mound measured roughly 250 square metres in area. There was a large stone surrounded by a distinct circle of smaller stones at the southern end. The weapons discovered prior to excavation were from the area south of the large stone. The remainder of the site consisted of a dense setting of stones in several layers. Between the stones, the soil was strongly mixed with soot and charcoal, in places to a depth of half a metre. Artefacts were found scattered among the stones and also in distinct clusters. The surface of the cemetery area was not distinct from its immediate surroundings, but in places in the bottom layer there were the remains of a perimeter laid with large stones. After the removal of the surface layer at the north end, a low cup-marked stone was discovered. At the north-east end, sloping towards the north and east, there were distinct, crossing plough-marks beneath the stone setting and in the north part of the cemetery an "offering pit", a kettle-shape depression (SÖYRINKI-HARMO 1996b, 102–118) full of crushed and charred seeds and grains (common wheat, emmer wheat, husked barley, rye, oats, peas, broad beans, flax and gold-of-pleasure) (AALTO 1996,

177–178). The funeral pyres are assumed to have been made at the highest point of the cemetery, north-east of the large stone (SÖYRINKI-HARMO 1996b, 118). Surface finds, some 30 m south of the cemetery, suggest an Iron Age house-floor (SAUKKONEN 1996, 28–29).

The finds from Vainionmäki include both male and female artefacts. Deriving from male burials were, among other objects, ten swords and sword fragments, seaxes, spearheads, shield bosses, bits, ornamental pins, tweezers and a razor. At least two of the swords bore Scandinavian animal ornamentation, having grips decorated in Salin´s style II, in addition to gilding and inlays of precious stones (SCHAUMAN-LÖNNQVIST 1996a, 53–62). The women's burials are represented by so-called crayfish brooches, even-armed brooches, an iron fibula with a band-shaped arch and hinges, a penannular brooch with rolled ends, fragments of a bracelet and a necklace, rings, beads and sections of chain ornaments (RANTA 1996, 36–42). Finds of tools and implements included knives, shears, a sickle, a billhook, chisels, a needle and rivets (SÖYRINKI-HARMO 1996a, 63–72). There were also four Arabian coins and their fragments, the oldest being a piece struck under Caliph Al-Walid between AD 705 and 715 (TALVIO 1996, 51–52). The oldest artefacts are from the middle of the seventh century, and the youngest from around AD 800, indicating that the cemetery was in use for a period of approximately 160 years (SCHAUMAN-LÖNNQVIST 1996b, 130–131).

Some of the objects had been in a fire, and some had been deliberately broken. Along with the weapons, ornaments and tools there were numerous pot sherds. Some of these had also been in the pyre, while others had probably been deposited in connection with the funeral feast or in sacrifices of food (HIRVILUOTO 1996, 73, 78–79). The finds of burnt bone include both human and animal bones, remains of canines, sheep, goat, bear and seal, and the finds of unburnt animal bones include remains of horse, cattle, seal and domestic pig. The unburnt teeth of horses and bovines are not from any cremation or burial-feast context but are evidence of some other aspect of the burial rite (FORMISTO 1996, 82–87).

The bone material suggests that at least 25 individuals were buried in the cemetery. Some of which were children, four of them 0–7 years old and four of them 0–14 years old. One of the children was not cremated (FORMISTO 1996, 82–83). Seven definite individual burials, all male, were investigated. Most of the weapons and other male objects formed distinct clusters, whereas women's artefacts were found scattered throughout the artefact layer. There was another distinct difference between menus and women's objects. Most of the women's objects were found in the highest part of the cemetery, to the north and north-east of the central stone, while most of the burials with the weapons were on the south-eastern or southerly slope of the site. In one case (graves 5 and 12) there were bones of male and female mixed with each other (HEIKKURINEN-MONTELL 1996, 88–101).

At the Vainionmäki site, the cemetery hill forms a concrete starting point when one seeks to analyse the burial pattern and its function as a reflector of the relationship between the here and the hereafter, in other words, between the living and the dead. This relationship is symbolised by, among other things, the stone circle that surrounds the cemetery and separates it from its surroundings, thus forming a border between the sacred and the profane. The association of ritual practices with the burial of the dead has been considered to indicate a belief in the continuation of life in some form even after death, a belief in the existence of an immortal soul. Because the dead were accompanied with weapons, implements, ornaments and food, it was believed that the afterlife was to some degree similar to the existence of the living.

In many cultures, the notion of the soul itself is pluralistic. A human being is thought to have several souls, at least one of which can join the company of the souls of previously dead ancestors when released from the dead body. In the afterworld, the individual is represented by the freesoul. This, however, can only leave the body after the latter has totally disintegrated or been transformed into another substance. Very small children were probably not thought to have souls (GRÄSLUND 1994, 17–18), which would explain the exceptional inhumation burial of the child at Vainionmäki. Among the Sami, the idea of a multi-part essence has prevailed until the present day. The free soul journeyed to the Realm of the Dead, where a new body was created for it. Human bones were regarded as being imbued with some kind of principle of life, and a possibility of returning to life was felt to last as long as the skeleton remained (PENTIKÄINEN 1995, 194–195).

Although it is difficult to interpret the meanings of prehistoric burial customs, there is literary evidence of the conceptual basis of cremation. One source is from Ancient Greece, from the time corresponding to Period V of the Scandinavian Bronze Age. The other is from Viking Age Russia. Book 23 of the *Iliad* tells how Patrocles appears in a dream to Achilles, bidding him as follows (SALO 1984, 141):

> "Bury me with all speed, that I pass the gates of Hades. Far off the spirits banish me, the phantoms of men outworn, nor suffer me to mingle with them beyond the River, but vainly I wander along the wide-gated dwelling of Hades. Now give me, I pray pitifully of thee, thy hand, for never more again

shall I come back from Hades, when ye have give me my due of fire." (from the *Iliad* 1966, 316 [translation by Ernest Myers]).

This implies that the purpose of cremation would have been to free the invisible spirit of the deceased, or the soul from the dead body. The lines of the *Iliad* also reveal another meaning attached to the burning of the dead. After cremation the soul will no longer come back to join the living; until then it can come to them.

The written source from AD 922/923 is an eyewitness account by the Arabian traveller Ibn Fadlan of the cremation of a Viking chieftain on the Volga in Russia. According to this account, a compatriot of the deceased who followed the ritual reveals the reason for it:

"Next to me was a Viking, whom I heard speaking to the interpreter who was with me. I asked the interpreter what the man had said, and he replied: 'You Arabs are stupid'. 'What do you mean?' I asked. 'Because you take those whom you love and respect most and lay them in the ground where the earth and the worms will eat them. We burn them in the blinking of an eye, and they step straight into paradise'." (SIMONSEN 1981, 59).

In both cases it is noted quite unequivocally that it is only the burning of the corpse that will free the soul of the deceased and let it proceed to Hades or to paradise. The mere burning the body was not, however, sufficient in itself to guarantee entry into the netherworld, and the Scandinavian sagas actually mention that a dog was needed to guide the soul on the dangerous journey to its final destination (DAVIDSON 1988, 57). Perhaps this is why many of the Vainionmäki Dead have been joined on the pyre by a dog (FORMISTO 1996, 84–87), a phenomenon that is also found in Swedish cemeteries (SIGVALLIUS 1994, 76–77). The dog is also a mythical animal in Finnish folklore, and it plays an important role in healing rituals, among other things (SIIKALA 1992, 101–103).

The idea of burning the corpse to free the soul reflects the individualistic or personal aspect of cremation. An evaluation of the relationship between this individualistic aspect and the communal or social aspect may be attempted through an interpretation of the meaning of the criss-cross ard-marks found in the north-eastern part of the cemetery, below the cemetery hill, the "offering pit" on their western border, and the cup-marked stone on the northern side of the cemetery. The ard-marks appear in a restricted area on a moraine slope falling off towards the north and north-west, so they cannot represent the remains of a field antedating the cemetery.

In Scandinavia, ard-marks were found in areas located downhill from Stone
Age, Bronze Age, and Iron Age cemeteries, and though some of them derive
from fields antedating the cemeteries, there is also evidence of ritual ploughing
(NIELSEN 1993, 166–167). In Russia, the first ard-marks under a burial mound
were discovered in 1991, albeit apparently belonging to an earlier field at the site
(NOSOV 1994, 60–62). The aforementioned "offering pit" containing whole and
crushed seeds and grains, is also connected with farming. A Finnish folk poem
concerned with cremation indicates that the cemetery was seen as a field while
the remains of the pyre, the burnt bones of the ancestors, were thought of as seeds
sown in the field (HIRVILUOTO 1987, 119–120; LEPPÄAHO 1950). The poem has
originally been linked with the Iron Age custom of burial in boats and reads, in
translation, as follows:

> "The Tarsian was burned
> In a copper boat
> An iron bottomed craft.
> Its embers were sown
> Into a field of no foundation,
> Into a ground of no substance."
>   (SKVR VII, 3, 738)

The cemetery linked the community's means of subsistence with past
generations. Farming had led to the adoption of a cyclical conception of time;
basing the calendar on the crop year. The latter, in turn, reflects the progress of
human life, from conception to birth, growth and death. Many religions, including
Christianity, place great importance on the fundamental idea that death is always
the prerequisite of a new life. Cup-marked offering stones have been found in the
vicinity of fields even where no cemeteries have been discovered. They are also
after prehistoric times connected to a fertility cult, and are known to have been
given offerings of milk and grain up to the nineteenth century (KIVIKOSKI 1966,
82).

As noted above, the burials on the cemetery hill of Vainionmäki are located
on the crown of the hill and on its southern and eastern slopes. The distribution
of pottery sherds, however, covers a much larger area, which suggests that the
dead received food offerings even after the funeral. Excavations have uncovered
evidence of an Iron Age settlement only 30 m south of the cemetery. If this
settlement represents the society that used the cemetery, the arrangement offered
excellent possibilities for intensive "communication" between current and former
members. The features of the cemetery reflecting the importance of farming and

their possible connection with the fertility cult and securing the continuation of the seasonal cycle can be interpreted as evidence of ancestor worship. According to old folk poems, the ancestors could be appealed to for help when danger threatened (HIRVILUOTO 1987, 123):

> Are there old folk,
> Long time resting,
> Weeks in the ground,
> Ages in the clay.
> May the swordsmen rise from the earth,
> From the clay with horses,
> To help the only son,
> To stand by the honoured man.
> A hundred men with swords,
> To fell the envious,
> To strike down all barriers.
> All the men from under the soil,
> From the great, black earth.
>   (SKVR VII, 5b, 18)

"Raising the Dead" is also mentioned in the sagas, where it is usually connected with eliciting information from the deceased (SIIKALA 1992, 184–186). However, this may be a much more fundamental phenomenon from cosmological point of view. The dead ancestors were really thought to "be there", as it were, and even to be in close contact with their living relatives. Thus, on the basis of the cult of the ancestors, there arose a conception of time where the Living and the Dead were together in the same moment, and where the past, the present, and the future co-existed simultaneously (JOHANSEN 1997, 29, 47).

Iron Age religion in Finland has been studied by Unto Salo. One aspect that he has suggested is the mythical union of Ukko, the god of thunder, and his woman in and the roll of thunder. This, in turn, is based on the theme of "the wedding of the gods" (hieros gamos). With reference to the fact that Finnish myths of the birth of fire describe Ukko as engendering fire, Salo regards the so-called oval fire-striking stones of the Early Iron Age as the archaeological expression of the myth. Ukko, in turn, has been associated with Thor, the Scandinavian god of thunder (SALO 1990a; SALO 1990b).

Fire was an important element in Iron Age burial ritual. It was with fire, i.e. in cremation, that the soul of the deceased was released and his personal status in the afterlife was ensured. Cremation also has a number of background factors related to the welfare of the whole community. The cemetery mound situated near the fields and dwellings provided a contact with the ancestors. The correct burial rituals and offerings of food ensured the favourable attitude of the ancestors towards present family and kin. Fire was also a necessary element when fields were cleared in the forest, preparing the soil to receive seed. The oval fire-striking stones, the ard-marks beneath burial cairns and the sacrificial stones in connection with the cairns are all tangible evidence of the connection between fire, ploughing, sowing and human reproduction in the minds of Iron Age people in Finland. To quote Mircea Eliade: "The assimilation of the sexual act to agricultural work is frequent in numerous cultures. In Satapatha Brahmana (VII, 2, 2, 5) the earth is assimilated to the female organ of regeneration (yoni) and the seed to the semen virile. 'Your women are your tilth, so come your tillage how you choose (ELIADE 1971, 26).'

The Iron Age worldview was characterised by concreteness and instrumental-ism. With the aid of the cemetery and the ancestors, the community would ensure crops and the success of animal husbandry and thus, indirectly, the subsistence of the present family and kin. Although originally intended to ensure the posi-tion of the individual in the afterlife, cremation, combined with the fertility rites of a farming culture developed a system of beliefs, which – like the shamanism of hunter-gatherers – placed the community before its individual members. This means that although farming, as a more developed means of livelihood, and the religion related to it were adopted from Scandinavia, the cult retained a great deal of old elements. As in the hunter-gatherer culture, where part of the catch was sacrificed to stone idols, the farmers now brought to the sacrificial stones grain and dairy products. But unlike shamanism, in which a specific animal was worshipped as the provider of food and welfare for the community, worship was now directed towards the ancestors as an instrument that was believed to aid in achieving the goal of good crops and success in raising animals.

Cremation burial gradually ceased after the adoption of the Christian faith, since it was considered a pagan practice. Instead, burning became an instrument of punishment and disgrace used against witches and heretics. The meanings given to cremation by the church had not totally disappeared even at the beginning of the twentieth century, when the public discussion on cremation was initiated in Finland. Looking back now, both the passionate tones of the discussion and the arguments presented by the various parties to bolster their standpoints seem remarkable. Both the arguments used by the opponents of cremation and those used by its reformist advocates were religious in nature, even though Protestant theology maintains that the treatment of the body has no effect whatsoever on the future of the deceased in the hereafter. The opponents charged that cremation would blaspheme the Christian belief in resurrection, while the reformers championed cremation specifically as an atheist form of burial even though the principal arguments had to do with general hygiene and the overcrowding of graveyards.

During the last few decades, cremation has increased in popularity and has also come to be seen as a neutral practice, unrelated to questions of religiousness or anti-religiousness. Nevertheless, among the various current burial practices it is specifically in cremation, primarily in the treatment of the ashes, where a variety of new options have become available, options that derive from individual preferences and which represent different meanings to different people. Such new options include scattering the ashes over the sea or in a special grove in the cemetery. Burial at sea may have to do with the deceased person's love for the sea due, to his or her profession or hobby, but it can also result from a wish not to have to take care of a cemetery plot.

On the other hand, many people place great store in having a locus where they can remember the deceased. An example of this, which is very popular nowadays, is the custom of visiting graves on Christmas Eve, the ancient Festival of the Dead. Although the small hearths in the Stone Age graves and the modern practice of setting candles on graves cannot be explained through a shared religious background or having the same symbolic meaning, both are however, connected with a family member who has passed into the hereafter. The division into proximate and distant netherworlds, with their respective cosmologies, is obviously no longer possible with our current varied selection of burial patterns. Thus, the next time a discussion arises concerning the justification of cremation, the question will no longer be a religious one – unless the accusation that cremation represents a wasteful use of energy is considered to represent a new kind of religion in the form of an ecological worldview.

Bibliography

AALTO, M. 1996
   Archaeobotanical samples. Appendix 6. *Vainionmäki – a Merovingian Period Cemetery in Laitila, Finland*. Helsinki, National Board of Antiquities, 6.

DAVIDSON, H. R. E. 1988
Myths and symbols in Pagan Europe. Early Scandinavian and Celtic religions.
New York, Syracuse.

EDGREN, T. 1992
Den förhistoriska tiden. Särtryck ur *Finlands historia* I. Ekenäs, 9–270.

ELIADE, M. 1971
The Myth of Eternal Return or Cosmos and History. *Bollingen Series* XLVI.
Princeton University Press.

FORMISTO, T. 1996
Osteological Analyses. *Vainionmäki – a Merovingian Period Cemetery in
Laitila, Finland.* Helsinki, National Board of Antiquities, 81–87.

GRÄSLUND, B. 1994
Prehistoric Soul Beliefs in Northern Europe. *Proceedings of the Prehistoric
Society* 60. London, 15–26.

HEIKKURINEN-MONTELL, T. 1996
Distribution of the archaeological material. *Vainionmäki – a Merovingian
Period Cemetery in Laitila, Finland.* Helsinki, National Board of Antiquities,
88–101.

HIRVILUOTO, A-L. 1987
Päättyvän rautakauden ja varhaiskeskiajan hautalöydöistä. *Muinaisrunot ja
todellisuus.* Jyväskylä. *Historian Aitta* XX, 119–128.

HIRVILUOTO, A-L. 1996
Pottery, burnt clay and slag. *Vainionmäki – a Merovingian Period Cemetery
in Laitila, Finland.* Helsinki, National Board of Antiquities, 73–80.

HOMEROS, 1966
Iliad. Prose translation by Ernest Myers.

JOHANSEN, B. O. 1997
Aspekter av tillvaro och landskap. *Stockholm Studies in Archaeology* 14.
Stockholm.

KIVIKOSKI, E. 1966
Suomen kiinteät muinaisjäännökset. *Tietolipas* 43. Helsinki.

LEPPÄAHO, J. 1950
Tarsilaisen poltto. *Kalevalaseuran vuosikirja* 30. Helsinki, 99–111.

NIELSEN, V. 1993
Jernalderens plöjning. *Store Vildmose. Vendsyssel Historiske Museum.* Hjörring.

NOSOV, E. N. 1994
New archaeological Data of the Economy of the Population of the Lake Ilmen Region in the Second Half of the First Millennium AD. *Fenno-ugri et slavi 1992. Prehistoric economy and means of livelihood. Museoviraston arkeologian osasto. Julkaisu* No 5. Helsinki, 60–63.

PENTIKÄINEN, J. 1995
Saamelaiset – Pohjoisen kansan mytologia. *Suomalaisen Kirjallisuuden Seuran Toimituksia* 596. Helsinki.

PURHONEN, P. 1996
Mortuary practices, religion and society. *Vainionmäki – a Merovingian Period Cemetery in Laitila, Finland.* Helsinki, National Board of Antiquities, 119–129.

PURHONEN, P. 1998
Kristinuskon saapumisesta Suomeen – Uskontoarkeologinen tutkimus. *Suomen Muinaismuistoyhdistyksen Aikakauskirja* 106. Helsinki.

RANTA, H. 1996
Personal ornaments. *Vainionmäki – a Merovingian Period Cemetery in Laitila, Finland.* Helsinki, National Board of Antiquities, 36–50.

RÄTY, J. 1995
The red ochre graves of Vaateranta in Taipalsaari. *Fennoscandia archaeologica* XII. Helsinki, The Archaeological Society of Finland, 161–172.

SALO, U. 1981
Satakunnan pronssikausi. *Satakunnan historia* I, 2. Rauma.

SALO, U. 1984
Pronssikausi ja rautakauden alku. Suomen esihistoria. *Suomen historia* I. Espoo, 98–249.

SALO, U. 1990a
Agricola´s Ukko in the Light of Archaeology. A Chronological and Interpretative Study of Ancient Finnish Religion. *Old Norse and Finnish Religions and Cultic Place-Names.* Åbo, 92–192.

SALO, U. 1990b
Fire-striking Implements of Iron and Finnish Myths Relating to the Birth of Fire. *Fenno-ugri et Slavi 1988. Iskos 9.* Helsinki, 49–61.

SAUKKONEN, J. 1996
The Vainionmäki site in Kodjala, Laitila. *Vainionmäki – a Merovingian Period Cemetery in Laitila, Finland.* Helsinki, National Board of Antiquities, 19–35

SCHAUMAN-LÖNNQVIST, M. 1996a
Weapons. *Vainionmäki – a Merovingian Period Cemetery in Laitila, Finland.* Helsinki, National Board of Antiquities, 53–62.

SCHAUMAN-LÖNNQVIST, M. 1996b
The Vainionmäki Society. *Vainionmäki – a Merovingian Period Cemetery in Laitila, Finland.* Helsinki, National Board of Antiquities, 130–135.

SIGVALLIUS, B. 1994
Funeral Pyres. Iron Age cremation in North Spånga. *Thesis and Papers in Osteology* 1. Stockholm.

SIIKALA, A-L. 1992
Suomalainen shamanismi – mielikuvien historiaa. *Suomalaisen Kirjallisuuden Seuran Toimituksia 565.* Helsinki.

SIMONSEN, J. B. 1981
Vikingerne ved Volga. Ibn Fadlans resebeskrivelse resumerat, deloversat og kommenterat af Jörgen Baek Simonse. Århus.

SKVR 1908–1948
Suomen kansan vanhat runot I–XIV. Helsinki.

SÖYRINKI-HARMO, L. 1996a
Tools and implements. *Vainionmäki – a Merovingian Period Cemetery in Laitila, Finland.* Helsinki, National Board of Antiquities, 62–72.

SÖYRINKI-HARMO, L. 1996b
The Formation of the Vainionmäki cemetery. *Vainionmäki – a Merovingian Period Cemetery in Laitila, Finland.* Helsinki, National Board of Antiquities, 102–118.

TALVIO, T. 1996

The coins. *Vainionmäki – a Merovingian Period Cemetery in Laitila, Finland.* Helsinki, National Board of Antiquities, 51–52.

VAINIONMÄKI 1996

Vainionmäki – a Merovingian Period Cemetery in Laitila, Finland. In: Purhonen, P. (ed.), *Helsinki, National Board of Antiquities.*

5

Great Rows of Fire: The Linearity of Hearths and Cooking Pits in Southern Scandinavia

RAIMOND THÖRN

"It is a matter of surpassing remark, when you come down to think about it, what a change in the landscape occurs when you have made a place of your own" (ROBERT ARDREY 1997, 353).

Introduction

Linearly distributed hearths and/or cooking-pits have been perceived at archaeological excavations during the last 30 years in northern Europe. At some locations these structures can be counted in hundreds and can therefore be described as "Great Rows of Fire". These structures seem to have such a spatial and chronological distribution, that they are to be connected with the South Scandinavian Bronze Age culture. Preliminary research has resulted in: a division of the linear assemblages into simple and complex systems; a tendency that the majority of the systems in question are to be seen in a period III–IV context and a tendency that the complex systems are older that than the simple ones. The question of the system's function is a rather elusive one, but in observing the overall tendency for repetition in the record, the hypothesis is that the systems should be regarded as products of ritual behaviour (HEIDELK-SCHACHT 1989; THÖRN 1992, 1993, 1996; LEVY 1982; COLPE 1970; RENFREW – BAHN 1991, 359).

At some locations hundreds of hearts and/or cooking pits without any linearly distribution have been observed during archaeological excavations. How are these clusters to be interpreted according to the Great Rows of Fire? It is in this respect important to focus on the topography, if one agrees with the thought that ideational elements have had a deep impact on the structuring of activities in the landscape. According to this three of four examples are given from the Malmö area in southern Sweden, where new results have emerged during the archaeological project "The Öresund Link Project" (In Swedish: Öresundsförbindelsen). The project was related to the infrastructural project involving building a bridge across The Sound between Sweden and Denmark and constructing about 25 km of highway around the city of Malmö. When the project is finished 29 reports dealing with results from 54 sites will be found in the bookshelves, together with eight syntheses dealing with scientific sub-projects. One of the four cases is from the island of Fyn in Denmark, where a site with about 1600 cooking pits in fifteen rows can be seen as the most extreme example that I have read about in my studies of this phenomenon.

Case studies

1. Fosie, Malmö, Sweden

The first case study consists of the lower portion of a slope that faces to the west. The area is situated between the plain in the west and the outer hummocky landscape in the east. Comprehensive remains of prehistoric settlements, which have been documented within the frames of large archaeological projects during the year 1979–1983 and 1998 (BJÖRHEM – SÄFVESTAD 1993; HADEVIK – GIDLÖF 2003), presumably have its explanation in the variation of the vicinity. The nearby brook has presumably been an important resource in this habitation.

A north by south oriented row with 13 cooking-pits has been documented (*Figure 5.1*). The approximately 0.5 to 1.0 metre big and approximately 0.5 metre deep structures have been dated with 95.4 % probability to the interval 1100–840 BC, i.e. to the early part of the Late Bronze Age.

An undated metalled road seems to tangent the row of cooking-pits. In regarding the surrounding area, one can say that it is surprisingly free from structures. Further up in the slope plenty of prehistoric settlements can be found, but these are dated to the Late Iron Age. The closest habitation from the Late Bronze Age can probably be found about 750 m to the east (BJÖRHEM – SÄFVESTAD 1993; HADEVIK – GIDLÖF 2003).

The area is situated between the plain in the west and the outer hummocky landscape in the east. In this borderland an important road has existed, according to a map from the later half of the 17th century. There are approximately 2500 years that seems to separate the cooking pits from the road. However, it shall though be taken into consideration that archaeological and culture geographical research has shown that the road very seldom has any conjunction with the villages from the

Late Iron Age. Furthermore, archaeological observations of the mounds from the Early Bronze Age have resulted in thoughts saying that the stretch of the road and the mounds surprisingly often seems to be found in each other's vicinity. These observations could be seen as arguments for a road with prehistoric ancestry (SKÖLD 1963; ERIKSON 2001; SAMUELSSON 2001). It shall though be said that such observations were published much earlier in Denmark. The Danish teacher H. C. STRANDGAARD observed a row of grave mounds in northern Jutland already in the 1880s (STRANDGAARD 1883). It was however the archaeologist Sophus Müller who in 1897 published the explanatory model suggesting that the grave mounds were located at roads, where also the Bronze Age settlements were to be found. It shall also be said that a metalled road dating from the Late Bronze Age have been documented in Malmö in 1998 (WINKLER 2001).

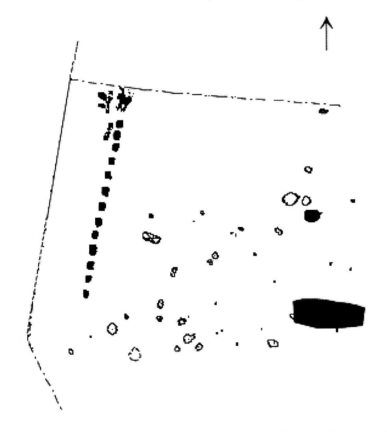

Figure 5.1. 13 cooking-pits and other structures from Fosie, Malmö, Sweden.
After HADEVIK – GIDLÖF 2003.

The archaeological excavations in the nearby parts of the western plain have mostly brought profane activities from the Early Iron Age into daylight. But on a distant part of a plateau, which was found at a distance of approximately 400 m as counted from the row of cooking pits, six graves collected within a small surface were found. Five of these were cremation burials, of which three can be dated to period III–IV of the Bronze Age, while the skeleton grave probably can be dated to the Roman Iron Age (SARNÄS – ENGSTRÖM, in press). But a connection between the row of cooking pits and other features should perhaps be searched for in another direction, because the row seems to point at the vicinity of an about 300 m to the north distant barrow of Early Bronze Age character named "Grötehög". About 70 m to the west of this barrow an archaeological excavation was carried out during the years 1985–1987. Among the features documented one can focus at a row of 24 postholes, which was oriented in WNW-ESE, i.e. in direction of the vicinity of the mound. In the west the about 40 m long row of post holes was finished by a pit, while the most eastern part probably is to be searched for outside the excavation area. In one of the postholes a quern stone was found which could be dated to the Late Iron Age (SAMUELSSON 1989). In the vicinity of the mound a burial field with 31 inhumation graves and about 160 cremation graves have been documented. Apart from one cremation burial dated to the Late Bronze Age, most of the burials can be considered as of Late Iron Age date (SAMUELSSON 2003). In the post holes eight flakes and a tool in preparation have been found. According to antiquarian Anders Högberg, Malmö, six of these flints can be dated to the Neolithic – Early Bronze Age and the tool in preparation can be dated to the Late Neolithic – the Early Bronze Age.[1] Adequate analogies to this row of post holes seem not to have been found in the literature concerned with the burial grounds of the Late Iron Age, why an older dating is probable. The flints found also indicate this. Immediately to the south of the barrow "Grötehög" an archaeological excavation that covered an area of 40 ha took place during the years 1979–1983. Among the large number of features found one can focus on a horseshoe-shaped post-construction with a diameter of 6 m and the open side turned to the WNW and the 760 m distant barrow "Grötehög". Two cremations, dated to the Late Bronze Age, were situated exactly on the line between the small house and the barrow. The feature is interpreted as a cult-house (BJÖRHEM – SÄFVESTAD 1993, 110 ff.).

---

[1]   I am grateful that my colleague Anders Högberg took some of his valuable time to help me analyse the flint material!

In the case study from Fosie it can be seen that a row of cooking pits appears to be in *isolation*, i.e. in a distant part of the area according to the longhouses from the Late Bronze Age. It is important to notice that the area is situated on a slope, i.e. an *exposed mode*. These two features can be found at the majority of the sites with rows of hearths or cooking pits, which have been studied in northern Europe (HEIDELK-SCHACHT 1989; THÖRN 1996). Relatively long distances between cooking pits and the habitation have occasionally been explained with a wish to minimize the threat of fire (LOMBORG 1977, 128), but if so, how shall we then explain the contemporary activities dealing with fire which actually have been documented inside or nearby the buildings of the Bronze Age? As an example can be mentioned the existence of cooking pits in a house from the Bronze Age in Thy, Denmark, (EARLE *et al*. 1998, 17 & fig. 4).

The structures with linearly distribution and the orientation towards a barrow of Early Bronze Age character, seems to indicate that the activities on this location should be seen as products of ritual behaviour.

When it goes for the 13 cooking pits in Fosie I propose that one has fully knowingly searched for this distant part of an area, which has been controlled by a community. One argument can have been that one has wanted to highlight its boundaries against settlements nearby. The message can have been that here have the forefathers lived during earlier generations, here we live now and here we want our children to dwell in the future, or as it has been suggested by BARRETT who states that: "The life history of settlement itself may have expressed the genealogy of humane existence (1997, 95)."

The borderland could have been attractive because of the ambiguity of being neither inside nor outside. It would then be easy to consider such a physical formation of the liminal as a location for activities associated with religion and rites of passage. Maybe has one prepared food for 13 people in these 13 in-a-row-laying cooking pits? Perhaps were these 13 persons earlier during the day attending a ceremony in order to bury their dear relative? If so, one can easily imagine that this human being has been present by words, thoughts and transactions. Maybe has the oldest inhabitant in the group talked about the dead person (HASTRUP – OVESEN 1982, 155), about events that have been and about coming actions beyond life? The present is mixed with the past. The narrator becomes the narrative (BLOCH 1974, 78). A "mixed time" is at hand, something that fits well in the zone between in- and outside and/or life and death.

2. Rønninge Søgård, Fyn, Denmark

The second case study is called Rønninge Søgård and is situated in the island of Fyn in Denmark. On a southern slope at Vindinge River the Danish archaeologists documented 302 cooking pits during the excavation seasons of 1972–1973. The site was found because of the exploitation of gravel, but unfortunately the exploiter did not inform the antiquaries every season. Because of this malfunction in the communications vast areas never became excavated by archaeologists, but HENRIK THRANE calculated in 1974 that there have been at least 500 cooking pits oriented from west to east in 15 rows. Because of later complementary archaeological excavations in 1975, this estimation was revised to about 1600 cooking pits (*Figure 5.2*). These features can be radiocarbon dated to the middle of the Bronze Age (1400–800 BC) and the finding of a pair of bronze tweezers indicates that we are at least dealing with period IV, i.e. 1100–900 BC. When it concerns with the function of the cooking pits at Rønninge Søgård, Thrane takes the standpoint that the place gives a profane impression. He leaves however the door open for a sacred interpretation when he writes that maybe the most distinguished votive object of the island of Fyn, from Mariesminde, is found 3.5 km to the west and that these eleven contemporary gold bowls scarcely are entirely without connections (THRANE 1974; 1989, 26).

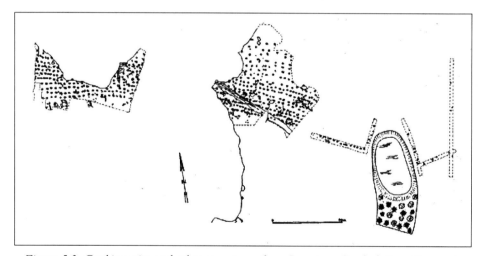

Figure 5.2. Cooking-pits and other structures from Rønninge Søgård, Fyn, Denmark. After THRANE 1989.

Rønninge Søgård is, with its at least 1600 cooking pits in fifteen rows, the most extreme site that I have read about in my studies of this phenomenon. When one thinks about this great effort of work, it seems fairly obvious that we have to do with a regional place of gathering. It also seems to be fairly obvious that all of the 1600 cooking pits have not been constructed and used during the same season or year. But to go from this simple saying to an exact account of the number of years that are involved is of course impossible, if one thinks about the chronological span that most often adheres with archaeological dating. After scrutinising the documentation material in the Danish archive in Odense, I think I can add something about the spatial distribution of the features though. I think that the cooking pits in the south have been dug with greater accuracy, than the northern ones. This could be interpreted as that the activities in the area started in the south.

The large amount of structures with linearly distribution seems to indicate that the activities on this location should be seen as products of ritual behaviour. I can only speculate in the question if the supposedly decreased symmetry between the rows should be explained with alterations in the ritual used or if a reduced intensity of the use of the area also led to lesser distinct traces. I shall say that we in the case of Rønninge Søgård can see the remains of a very clear willpower, which belonged to one or several people (THÖRN in press).

3. Södra Sallerup, Malmö, Sweden

The third case study deals with an area at the upper portion of a slope that faces eastward. Here the topography can be classified as a hummocky landscape, which means an environment consisting of heights of varying size in between of swampy areas. The moraine in the vicinity contains big blocks of writing chalk, which is unique in a Swedish perspective. The flint which can be found inside the chalk is of very high quality and it was mined already 6000 years ago, i.e. during the oldest phase of the Neolithic. In present time an industrial exploitation of writing chalk has been going on until some years ago. This has give rise to a rather large number of archaeological investigations during the seventies and the eighties. The archaeological results have shown that the vicinity has been used intensively during the Stone- as well as the Bronze and Iron Ages (NIELSEN – RUDEBECK 1991).

On the upper part of the eastern slope eleven structures could be found, which could be classified as hearths or cooking pits. Dating by Carbon 14 have been performed on seven of the structures and the results can be found within the interval 1300–1000 BC, i.e. the Bronze Age periods III and IV. The eleven structures are situated between a habitation approximately 70 m to the west and a culture layer approximately 10 m to the east. Both settlement and culture layer are to be found in a Middle Bronze Age context (THÖRN 1995).

When studying the layout of the structures it is rather clear that the structures within this site seems to be grouped in two clusters and therefore can not be of the same category as the two earlier examples (*Figure 5.3*). The western cluster contains six structures and the eastern one five. Perhaps the eye of the beholder wants to see it as though the clusters have a NNW to SSE and a respectively northwest to south-east orientation, but probably these are pseudo-observations. A more "objective" observation is that three hearths in the western cluster were fairly equally constituted, which among other characteristics can be shown by the fact that they all had volumes of 40 litres each. This could then indicate that at least half of the structures in the western cluster have been constructed at the same opportunity.

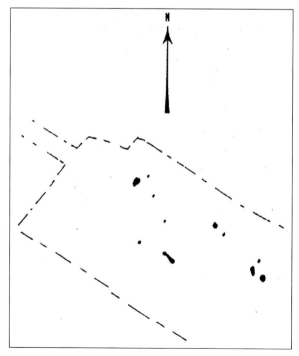

Figure 5.3. Cooking-pits and other structures from Södra Sallerup, Malmö, Sweden. After THÖRN 1995.

The structures can to some extent be looked upon as being constructed in pairs. An archaeo-botanical analysis, undertaken by Stefan Gustafsson at the Environmental Archaeology laboratory at Umeå University, have shown the existence of emmer (*Triticum dicoccum*) from the two westernmost structures (132 and 133) in the western cluster and from the two most easterly structures in the eastern cluster (140 and 142) (GUSTAFSSON 1995). Gustafsson interprets the structures as that they have been used for roasting of emmer. This hulled wheat is very hard to peel and it is almost needed that it shall be roasted in order to be able to remove the chaff. Apart from that the roasting enables peeling, it also makes the grains tastier and more lasting when stored. Gustafsson is also of the opinion that the cooking pits have been used as roasting pits, while one can have burnt the chaff. This fits very well with the fact that the western pair of structures consists of one hearth and one cooking pit. The same allotment can also be found at the eastern pair.

The structures without linearly distribution, the relatively nearness to a settlement and the existence of roasting pits, seems to indicate that the activities on this location should be seen as profane in a household context.

4. Västergård, Malmö, Sweden

The fourth case study is situated on a striking hill in the hummocky landscape. The hill had almost entirely come to perish by the exploitation of gravel, but its loamy border had not become destroyed and here could 129 cooking pits and 27 hearths be documented (*Figure 5.4*). The nine radiocarbon dating gave an interval of dating from 1520 BC to 480 BC (cal. 2 σ), but the centre of gravity seems to be found from 1200 to 1100 BC, i.e. the middle of the Bronze Age. A cluster in the southeast was already observed during the trial excavation and it was also noticed that the structures was to be found immediately up to a formerly swamp. After the main excavation a picture emerged where the cooking pits were circumscribed by four small marshes (ÖIJEBERG 1999, 94 ff. and appendix 18:2; GREHN *in press*). About 600 m to the northeast of these clusters one have documented a longhouse from the middle of the Bronze Age and graves from both the Early and Late Bronze Age (ÖIJEBERG 1999, 101f.; GREHN *in press*).

The structures without linearly distribution, the relatively nearness to both a settlement and graves, seems to indicate that the activities on this location could be seen as profane in a household context or as products of ritual behaviour. It can maybe seem to be tough nowadays to walk 600 m with heavy burdens in order to cook ones food, but probably they did not think that way during the Bronze Age. I would hold for likely that at least the cooking pits on the east portion of the hill have been used by the people that lived on the eastern settlement. Although we have conducted archaeological investigations on vast areas in the vicinity during later years, there will of course still exist white spots on the archaeological map. It would not be surprising to find out that the westernmost cooking pits belonged to a settlement that was located to the west of the hill, but future archaeological investigations may verify or falsify this.

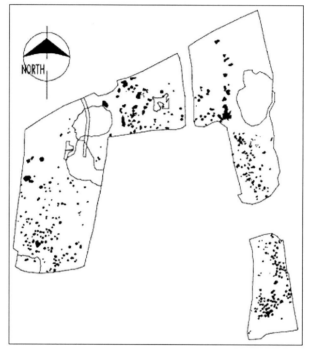

Figure 5.4. Cooking-pits and other structures from Västergård, Malmö, Sweden. After GREHN in press.

Concluding Remarks

Fire has been used in the South Scandinavian Bronze Age Culture to different extent and in different contexts. Thoughts about how the fire should be used have led to different placement in the landscape. In one case study we can see the traces after how one have roasted the grain near the home, in another example one has visited a somewhat distant hill almost surrounded by water in order to make one's food in cooking pits.

I can see products of a controlling and a rectifying factor in the archaeological record, which could be searched for in the religious sphere. I can then find support in similar opinions from scientists, where the fundamental idea may be expressed as by Dr. JANET E. LEVY in the following quotation: "Because religious activity is formal and repetitive, its material remains will tend to be repetitive in the archaeological record" (1982, 117). Fire protects and purifies and these abilities have probably been seen as valuable when great rows of fire have highlighted the limit between life and death.

The examples presented show us that a cooking pit is not always only a cooking pit. It shall be placed in a larger context and then we sometimes can see differences in function between different sites. But the function is still a rather elusive one, to a certain degree because of the problem to catch the sometimes subtle signals of what was sacred and what was profane. To understand and trying to grasp events now and then, one must categorize, typologize and generalize. The result can be that difficulties and what is hard to understand disappears and, more or less, clear and homogenous structures appears. One has then achieved something because one can now study the events and interpret them. But one must also be aware of the fact that one at the same time also has lost the ephemeral, the problematic and the heterogeneous. That seems to be the case with the dichotomy sacred and profane. At a conference in Oxford 1989 JAN HARDING asked the question: "…are we justified in assuming that the spatial separation of monuments and living areas equates to a division of social practices into ritual and domestic?" (HARDING 1991, 146). Harding fetched "the answer" from an article by Paul Lane where he in the introduction is explaining that this dichotomy: "…are a useful short-hand for defining the dominant characteristics of particular entities, their use introduces a set of largely ethnocentric, and frequently androcentric, assumptions, which serve to reinforce and reproduce an appearance of mutual exclusiveness and opposition between these two aspects of human action" (LANE 1986, 181). Later in the text Lane is developing his thoughts and he puts forward "that an analogous dichotomy reinforces the assumption that domestic functions are somehow universal" but "the empirical existence of private and public spheres, domestic and ritual contexts must not be taken as self-evident" (LANE 1986, 183 f.). I think that it is important that Lane is looking at "sacred and profane" as aspects of human action, which gives a more nuance way of looking at things. Lane is citing the social anthropologist GILBERT LEWIS, who is using results from the Norwegian anthropologist FREDRIK BARTH's studies of nomads in southern Iran (LEWIS 1980, 15; BARTH 1961). Barth gives examples of that one performance can contain both aspects. Harding could have used Maurice Bloch instead of citing Lane. At a conference in 1982 BLOCH reminded the audience that "Religion, however, in non-industrialised society itself proves to be a problem as it cannot be identified empirically since most of these societies do not have a sphere of life clearly marked out as religion. As anthropologists are fond of saying, religion is embedded". Bloch suggests though that religion can be identified by studies of ritual (BLOCH 1985, 34). But when studying supposedly remains of prehistoric rituals one must have in mind that religion is not equal with ritual, or as EDWARD SHILS has expressed it: "beliefs could exist without rituals; rituals, however, could not exist without beliefs" (SHILS 1968, 736).

So, networks of complex meaning are imposed upon the prehistoric landscape and it is up to us archaeologists to make the interpretations. Topographical observations, and sometimes archaeo-botanical analyses, give us important information that indicates function and context. But the question of the linearly distributed cooking pits' function is still a rather elusive one, but in observing the overall tendency of repetition in the record, the hypothesis is that the systems should be regarded as products of ritual behaviour.

Bibliography

ARDREY, R. 1997 [1967]
  The territorial imperative: a personal inquiry into the animal origins of property and nations. New York, Tokyo, London, Kodansha International.

BARRETT, J. C. 1997 [1994]
  Defining domestic space in the Bronze Age of southern Britain. In: Parker Pearson, M. – Richards, C. (eds.), *Architecture and Order. Approaches to Social Space.* London, New York, Routledge, 87–97.

BARTH, F. 1961
  Nomads of South Persia. Oslo, Oslo University Press.

BJÖRHEM, N. – SÄFVESTAD, U. 1993
  Fosie IV. Bebyggelsen under brons- och järnålder. Malmöfynd 6. Malmö, Malmö Museer.

BLOCH, M. 1974

Symbols, Song, Dance and Features of Articulation. Is religion an extreme form of traditional authority? *Archives Européennes de Sociologie* Tome XV, Numéro I, 55–81.

BLOCH, M. 1985

From cognition to ideology. In: Richard Fardon (ed.), *Power and knowledge. Anthropological and sociological approaches. Proceedings of a conference held at the University of St. Andrews in December 1982.* Edinburgh, Scottish Academic Press, 21–48.

COLPE, C. 1970

Theoretische Möglichkeiten zur Identifizierung von Heiligtümern und Interpretation von Opfern in ur- und parahistorischen Epochen. In: *Vorgeschichtliche Heiligtümer und Opferplätze in Mittel- und Nordeuropa: Bericht über ein Symposium in Reinhausen bei Göttingen in der Zeit vom 14. bis 16. Oktober 1968.* Herausgegeben von Herbert Jankuhn. Abhandlungen der Akademie der Wissenschaften in Göttingen. Philologisch-Historische Klasse. Dritte Folge, No. 74. Göttingen, 18–39.

EARLE, T. – BECH, J. H. – KRISTIANSEN, K. – APERLO, P. – KELERTAS, K. – STEINBERG, J. 1998

The Political Economy of Late Neolithic and Early Bronze Age Society: the Thy Archaeological Project. *Norwegian Archaeological Review* Vol. 31, No. 1, 1–28.

ERIKSON, M. 2001

En väg till Uppåkra. In: Larsson, L. (ed.), *Uppåkra. Centrum i analys och rapport.* Acta Archaeologica Lundensia Series 8, No. 36. Uppåkrastudier 4, 167–176.

GREHN, F. In press

Öresundsförbindelsen. Västergård 18. Rapport över arkeologisk slutundersökning. Rapport nr 33. Malmö, Malmö Kulturmiljö.

GUSTAFSSON, S. 1995

Sallerup MHM 7832. Arkeobotanisk analys av förkolnade växtrester från härdar och kokgropar i Södra Sallerup, Skåne. Umeå, Umeå Universitet, Miljöarkeologiska laboratoriet. Unpublished.

HADEVIK, C. – GIDLÖF, K. 2003
Öresundsförbindelsen Fosie 11AD samt Broläge Larsbovägen. Rapport över arkeologisk slutundersökning. Malmö, Malmö Kulturmiljö.

HARDING, J. 1991
Using the unique as the typical: monuments and the ritual landscape. In: Garwood, P., Jennings, D., Skeates, R. – Toms, J. (eds.), *Sacred and profane. Proceedings of a conference on archaeology, ritual and religion.* Oxford, Oxford University Committee for Archaeology, Monograph No. 32, 141–151.

HASTRUP, K. – OVESEN, J. 1982
Främmande kulturer. Den moderna etnografins grunder. Södertälje, Gidlunds.

HEIDELK-SCHACHT, S. 1989
Jungbronzezeitliche und früheisenzeitliche Kultfeuerplätze im Norden der DDR. In: *Religion und Kult in ur- und frühgeschichtlicher Zeit. XIII. Tagung der Fachgruppe Ur- und Frühgeschichte vom 4. bis 6. November 1985 in Halle (Saale).* Im Auftrag der Historiker-Gesellschaft der DDR herausgegeben von Friedrich Schlette und Dieter Kaufmann. Berlin, Akademie-Verlag, 225–240.

LANE, P. 1986
Past practices in the ritual present: examples from the Welsh Bronze Age. *Archaeological Review from Cambridge* 5(2), 181–192.

LEVY, J. E. 1982
Social and Religious Organization in Bronze Age Denmark. An Analysis of Ritual Hoard Finds. British Archaeological Reports (BAR). International Series 124.

LEWIS, G. 1980
Day of shining red. An essay on understanding ritual. Cambridge, Cambridge University Press.

LOMBORG, E. 1977
Bronzealderbopladsen på Skamlebæk radiostation. In: Albrethsen, S. E. (ed.), *Antikvariska studier tilegnet Knud Thorvildsen på 70-årsdagen 18. december 1977.* Fortidsminder 7. København, Fredningsstyrelsen, 123–130.

MÜLLER, S. 1897
Vor oldtid. Danmarks forhistoriske archæologi. København, Det nordiske forl.

NIELSEN, B. – RUDEBECK, E. 1991
Introduktion till arkeologi i Södra Sallerup. En översikt över utgrävningarna kring Ängdala gård. *Elbogen. Malmö Fornminnesförenings Årsskrift 1990.* Årgång 58, 64–97.

ÖIJEBERG, J. 1999
Kommungränsen – Riksväg 11. In: Billberg, I. – Magnusson Staaf, B. (eds.), *Öresundsförbindelsen II. Rapport över arkeologiska förundersökningar.* Stadsantikvariska avdelningen, Kultur Malmö, 94–105.

RENFREW, C. – BAHN, P. 1991
Archaeology. Theories, Methods and Practice. London, Thames & Hudson.

SAMUELSSON, B. - Å. 1989
Rapport. Arkeologisk huvudundersökning 1985–87. Kv. Bronsdolken 7, Lockarp sn, RAÄ 64, Malmö kn, M-län, Sk. Unpublished.

SAMUELSSON, B. - Å. 2001
Kan gravar spegla vägars ålder och betydelse? Ett exempel från Söderslätt i Skåne. In: Lars Larsson (ed.), *Uppåkra. Centrum i analys och rapport.* Acta Archaeologica Lundensia Series in 8°, No. 36. Uppåkrastudier 4, 177–184.

SAMUELSSON, B. - Å. 2003
Ljungbacka: a Late Iron Age cemetery in south-west Scania. *Lund Archaeological Review* 7/2001, 89–108.

SARNÄS, A. – ENGSTRÖM, T. In Press
Öresundsförbindelsen. Lockarp 7H & Bageritomten. Rapport över arkeologisk slutundersökning. Rapport nr 19. Malmö, Malmö Kulturmiljö.

SHILS, E. 1968
Ritual and crisis. In: Cutler, D. R. (ed.), *The Religious Situation.* Boston: Beacon Press, 733–49.

SKÖLD, P. E. 1963
En väg och en bygd i gammal tid. *Ale* 2/1963. Historisk tidskrift för Skåne, Halland och Blekinge, 1–15.

STRANDGAARD, H. C. 1883
Om en mærkelig række af høje i det vestlige Jylland. *Samlinger til Jydsk Historie og Topografi* 1882–1883. IX bind, 88–92.

THRANE, H. 1974
Hundredvis af energikilder fra yngre broncealder. *Fynske Minder 1974 tilegnet Erling Albrectsen på 70 års dagen.* Odense, Odense Bys Museer, 96–114.

THRANE, H. 1989
De 11 guldskåle fra Mariesminde: vidnesbyrd om en broncealderhelligdom? *Fynske Minder* 1989, 13–30.

THÖRN, R. 1992
Käglinge grustäkt – aspekter kring ett komplext boplats- och kokgropsområde. In: Ödman, C. (ed.), *Arkeologi i Malmö. En presentation av ett antal undersökningar utförda under 1980-talet.* Rapport nr 4. Malmö, Malmö Museer, Stadsantikvariska avdelningen, 9–36.

THÖRN, R. 1993
Eldstadssystem – fysiska spår av bronsålderskult: ett försök att spåra kultplatser och kulturinfluenser. C-uppsats i arkeologi ht 1993. Lund, Arkeologiska institutionen, Lunds Universitet. Unpublished.

THÖRN, R. 1995
Sallerupsvägen: delen väster om Särslövsvägen. Rapport över arkeologisk slutundersökning. Malmö, Stadsantikvariska avdelningen, Malmö Museer. Unpublished.

THÖRN, R. 1996
Rituella eldar: linjära, konkava och konvexa spår efter ritualer inom nord- och centraleuropeiska brons- och järnålderskulturer. In: Engdahl, K. – Kaliff, A. (eds.), *Religion från stenålder till medeltid. Artiklar baserade på Religionsarkeologiska nätverksgruppens konferens på Lövstadbruk den 1–3 december 1995.* Linköping: Riksantikvarieämbetet. Arkeologiska undersökningar. Skrifter nr 19, 135–148.

THÖRN, R. In press
Kokgropsrelationer. Oslo, Universitetets Kulturhistoriske Museer.

WINKLER, M. 2001
Vägen från Södra Sallerup. In: Lars Larsson (ed.), *Kommunikation i tid och rum.* Report Series No. 82. University of Lund, Institute of Archaeology, 41–49.

6

Lighting-up the End of the Passage:
The Way Megalithic Art Was Viewed and Experienced

GEORGE NASH

Introduction

...he came to a mouth of a passage covered with a square stone similar to that at [nearby] Plasnewydd, anxious to reap the fruits of his discovery he procured a light and crept forward on his hands and knees along the dreary vault, when lo! In a chamber at the further end a figure in white seemed to forbid his approach. The poor man had scarcely power sufficient to crawl backwards out of this den of spirits... (REVEREND JOHN SKINNER 1802)

There is a limited but significant passage-grave art tradition in England and Wales which, although restricted to two passage grave monuments in Anglesey, Bryn Celli Ddu and Barclodiad y Gawres and a destroyed megalithic structure located in a park in Liverpool, the Calderstones, marks the eastern extent of megalithic art in Britain and Ireland. The passage grave tradition is also one of the last megalithic architectural styles of the Neolithic. Outside these two areas are a number of sites that possess simple decoration, usually restricted to single and multiple cupmarks (DARVILL – WAINWRIGHT 2003; SHARKEY 2004; NASH *et al.* 2006) (see also *Table 6.1*). However, these monuments are not classified as passage graves.[1] Outside England and Wales; in Ireland and north-western France, megalithic art and passage grave architecture is both numerous and complex in form and many of the traits incorporated into these monuments are also replicated in the three British examples (FORDE-JOHNSON 1956; LYNCH 1967, 1970; SHEE-TWOHIG 1981).

These three British passage graves each possess complex carved art that is usually located within the inner section of the passage or forms part of the chamber

---

[1] FRANCES LYNCH however does describe several of these as being 'short passage graves (1969a; 1970)'. However, I am inclined to suggest that these monuments have little architectural or chronological associations with the passage grave tradition *per se* of the Late Neolithic.

alignment (NASH *et al.* 2006). Both Anglesey monuments have the passage and chamber architecture incorporated into large covering mounds. Despite the intense and comprehensive archaeological investigations of these monuments and the near complete destruction of the Calderstones monument, one can make an assessment of how these monuments may have functioned during ceremonial and ritual-symbolic events. Recent investigations by the author have shown how light may have played an important role especially in illumination of various parts of the chamber and passage architecture; more significantly, how the rock-art was viewed and experienced.

Table 6.1. Megalithic chambered monuments with rock-art in wales and the border counties.

Site	Grid ref.	Art	Location	References[2]
Arthur's Stone, Herefordshire	SO 318 431	Cupmarks	Portal stone	CHILDREN – NASH 1994; 2006; HEMP 1935
Bachwen, Caenarvonshire	SH 407 495	Cupmarks	Capstone	DANIEL 1950; HEMP 1926; LYNCH 1969a
Barclodiad y Gawres, Anglesey	SH 329 707	Spirals, chevrons, zigzags, lines, lozenges, cupmarks	Chamber	POWELL – DANIEL 1956; LYNCH 1969a; NASH *et al.* 2006; SHEE-TWOHIG 1981
Bryn Celli Ddu, Anglesey	SH 508 702	Serpentine, spiral, cupmarks	Chamber, rock outcropping	DANIEL 1950; HEMP 1930; LYNCH 1969a; NASH 2006; NASH *et al.* 2006; SHEE-TWOHIG 1981
Calderstones, Liverpool	SJ 405 875	Concentric circles, cupmarks, footprints, lines/grooves, spirals,	Chamber uprights (destroyed passage grave)	DANIEL 1950; FORDE-JOHNSTON 1956; NASH 2006; SHEE-TWOHIG 1981
Carreg Coetan Arthur, Pembrokeshire	SN 061 359	Cupmarks?	Capstone	CHILDREN – NASH 1997; NASH 2006
Cerrig y Gof, Pembrokeshire	SN 037 389	Cupmarks	Capstones, rock outcropping	NASH 2006
Cromlech Farm, Anglesey	SH 360 920	Cupmarks, horse-shoe carving	Monument architecture and rock outcropping	NASH *et al.* (2006)
Cist Cerrig, Caenarvonshire[3]	SH 543 384	Cupmarks	Rock outcropping	LYNCH 1969a
Cae Dyni, Caenarvonshire	SH 511 382	Cupmarks	Located on two uprights	NASH *et al.* (2006)

---

[2]   References refer to discussions specific to rock-art rather than the monument.
[3]   Also known as Treflys.

Dyffryn Ardudwy, Merioneth	SH 588 229	Cupmarks	North portal of the western chamber	POWELL 1973; SHARKEY 2004
Garn Turne, Pembrokeshire	SM 979 272	Cup-and-ring, Cupmarks	Capstone, rock outcropping	NASH *et al.* (2006)
Garn Wen Cemetery, Pembrokeshire	SM 948 390	Cupmarks	Rock outcropping	NASH *et al.* (2006)
Llannerch,	SH 559 379	Cupmarks	Remains of chambered tomb?	SHARKEY 2004
Morfa Bychan, Carmarthenshire	SN 221 075	Cupmarks	Rock outcropping	SHARKEY 2004
Pentre Ifan, Pembrokeshire	SN 099 370	Cupmark, spiral?	Portal stone	LYNCH 1972
Treflys, Caenarvonshire	SH 543 384	Cupmarks	Rock outcropping	BARKER 1992; HEMP 1938
Trellyffaint, Pembrokeshire	SN 082 425	Cupmarks	Capstone	BARKER 1992; CHILDREN – NASH 1997
Ty Illtud, Breconshire	SO 098 263	Geometric forms, semi- representative figures (medieval)	Chamber uprights	CHILDREN – NASH 2001; RCAHMW 1997
Ty Newydd, Anglesey	SH 617 112	Cupmarks	Capstone	SHARKEY 2004

This chapter will assess the impact of light, in particular fire on the chamber, façade and passage areas, focusing on how and why megalithic rock-art was illuminated. I suggest that megalithic art was deliberately placed in such a way as to visually change the ambience of the space between various sections of the passage and chamber, in particular at the point where natural light in the passage area fades and was no doubt was replaced by the illumination of fire. Based on two types of light and their intensity, two cognitive emotions are at play. One based on the familiarity and safety of the façade and the outer passage areas, the other based on the ignorance and anticipation of the inner passage and chamber which are initially dark, foreboding and, to many, unknown.

The Data Set

The distribution of the passage grave tradition is within the Atlantic zone of Europe, occupying five major core areas, from the Iberian Peninsula to southern Scandinavia. Its development can be clearly traced and it probably terminates in Wales and north-west England by about the mid-3rd Millennium BC (SHEE-TWOHIG 1981). The majority of the passage graves, in particular those monuments found in southern Scandinavia, are not decorated with megalithic art and are considered earlier than those that are, dating between 3,200 and 3,500 cal. BC (TILLEY 1991, 77) and probably represent an initial wave of immigrant farming or the translocation of ideas (or both). Megalithic art can therefore be considered a secondary tradition that occurs only in certain passage grave areas such as the Iberian Peninsula, Brittany, Ireland, Wales and Scotland. The latter two areas are to no more than 12 sites with megalithic art suggesting limited and late contact with area that were fully immersed in the passage grave tradition.[4]

Discussing the distribution of megalithic rock-art in Britain and Ireland is not new. Significant work has been undertaken in Ireland by a number of eminent archaeologists such as MICHAEL O'KELLY (1982), GEORGE EOGAN (1986), MUIRIS O'SULLIVAN (1986; 1993), ELIZABETH SHEE-TWOHIG (1981) and a comprehensive summary by GABRIAL COONEY (2000). The passage graves of Newgrange, Knowth and Dowth, located within the Boyne Valley, and Fourknocks and Loughcrew (Sliabh na Callighe), in County Meath all possess significant megalithic rock-art (SHEE-TWOHIG 1981, COONEY 2000). Some designs arguably originate from the south, along the Atlantic seaboard, within the core areas of Brittany and, according to EOGAN, from the Iberian Peninsula as well (1986, 172). Based on a limited but significant radiocarbon dating range, one can trace the development of the passage grave tradition. It was generally considered that the movement is from south to north with its zenith in central and southern coastal Ireland during the latter part of the 4th millennium BC (*Table 6.2*).

The demise of the passage grave tradition, in particular those monuments that possess rock-art, appears to occur in North Wales and north-west England, although up to seven sites exist in Scotland and Orkney (PIGGOT 1954; SHEE-TWOHIG 1981). Simple passage graves located on the Isles of Scilly and referred to as entrance tombs date to the late Neolithic but have no megalithic art. It is probable that the same architectural influences were moving northwards but some communities may have considered megalithic art time consuming and developed new ways of expressing burial rites, maybe through the production and use of portable artifacts such as pottery.

---

[4]   SHEE-TWOHIG (1981, 93) notes 11 passage graves in Wales and around 250 in Scotland.

Table 6.2. Radiocarbon date ranges for passage grave construction along the Atlantic Seaboard[5]

Site Name	Sample/Area	*RC[14] date*	Lab. No.	Reference
Alberite Cadiz, Spain	Funerary phase	5320±70 BP **4255–4000 cal. BC**	Beta-80602	RAMOS MUNOZ – GILES PACHECO 1996
El Palomar Cadiz, Spain	Funerary phase	4930±70 BP **3780–3640 cal. BC**	Beta-75067	RAMOS MUNOZ – GILES PACHECO 1996
Ile Gaignog, Brittany	Construction phase (tomb C)	3850±300 bc **4630 cal. BC**	Gif – 165	SHEE-TWOHIG (1981, 51)
Barnenez, Brittany	Sealed deposit in chamber G	3050±150 bc **3800 cal. BC**	Gif – 1309	SHEE-TWOHIG (1981, 51)
Barnenez, Brittany	Lower deposit in chamber F	3000±150 bc **3600 cal. BC**	Gif – 1556	SHEE-TWOHIG (1981, 51)
La Hougue Bie, Jersey	Primary cairn deposit	3450–3200 bc **4365-4055 cal. BC**	Beta-77360/ ETH-13185)	PATTON (1995, 582)
Knowth, Ireland	Monument construction – charcoal from mound	2455±35 bc **3100 cal. BC**	GrN-12357	EOGAN (1986, 225)
Knowth, Ireland	Pre-mound surface, con-temporary with construc-tion of the mound	2540±60 bc **3150 cal. BC**	GrN-12358	EOGAN (1986, 225)
Newgrange, Ireland	Monument construction -burnt soil putty used to infill the cracks between the roofing slabs	2475±45 bc **3100 cal. BC**	GrN-5462-C	O'KELLY (1982, 230–1)
Newgrange	Same as above	2465±40 bc **3100 cal. BC**	GrN-5463	O'KELLY (1982, 230–1)

Recently, in 2005, part of an upright from a nearby destroyed passage grave has been discovered in a Bronze Age barrow at Balblair, Inverness, forming one of the cist walls (DUTTON – CLAPPERTON 2005). This carving, present on a former upright, shows a new and altogether unique curvilinear design that is not found on any other megalithic stone in the British Isles. The stone, forming the wall of a Bronze Age cist probably originated from a nearby destroyed passage grave.[6] The stone decoration comprises two large gouges, possibly cupmarks with each cup measuring around 0.20 m in diameter and up to three finely carved lines radiating away from one of the cups. One is perforated and may have possessed a functional rather than a decorative role. The second and more intriguing design field is located on the upper section of the slab comprising of a series deeply gouged semi-ovate lines which resemble (but not necessarily is) the trunk and branches

---

[5] There are no radiocarbon dates for the Welsh passage-graves.

[6] The reuse of stone at sites such as Balblair is significant but is not discussed here.

of a tree. However, equally probable and incorporating the two large gouged cups
from the lower section of the slab, the two design fields may represent a stylised
penis (with accompanying testes).[7] The upper design, measuring approximately
0.70 m x 0.70 m is symmetrically placed with the outer gouges running parallel
with a possible deliberately cut and shaped recess. Despite the uniqueness of this
art, however, elsewhere in Scotland the designs show evidence for a regional
style developed from probable contact and exchange with monument builders
in Ireland. Spirals and concentric circles at Pickaquoy and Eday Manse, both in
Orkney, have similar decorative styles with monuments in central Ireland (SHEE-
TWOHIG 1981, 136–137).

The people who constructed the monuments in Anglesey and on Merseyside
would have been involved in varying degrees of socio-political contact and
exchange with passage grave builders in central eastern Ireland, operating in
what HERITY and others refer to as the *Irish Sea Zone* (1970, 30–33). This zone
is identified mainly through the stylistic similarities in monument building and
the deposition of material culture within each of the core areas that lie within this
zone. It is clear that the architectural traits that form the passage grave tradition
within both Ireland and North Wales are similar.[8]

Excavations undertaken at Newgrange and Knowth in the Boyne Valley,
Ireland, have revealed that megalithic art is not just confined to the passage and
chamber areas. Both monuments, two of the largest passage graves in Europe,
possess rock-art on the front and back of kerbstones which delineate the extent
of each of the mounds (EOGAN 1986). Although specific groups of carvings are
used, no two stones are the same.[9]

---

[7] Sexual and body iconography is not uncommon in early and later prehistoric art. In
     megalithic art THOMAS – TILLEY (1993) for example has suggested that certain infilled
     rectangular designs represent the rib cage (and acts as a metaphor for the 'body' of the
     tomb).

[8] Earlier architectural traditions in Ireland and Western Britain are also present such as
     the wedge tombs of the Carlingford-Clyde Group and the Court tombs of southern
     Ireland and Cornwall *etc.* (DANIEL 1950).

[9] Interestingly, recent experimentation with the 3D Archaeology Society at nearby
     Dowth suggests that the only way some of the more hidden designs can be read , i.e.
     those found on the sides and rear of the kerbstones, are best seen at night and using an
     artificial light source, as if Neolithic people were using the site at night, by using fire.

At Knowth, EOGAN recognised two distinct forms of megalithic art, one angular, the other curvilinear (1986).[10] Both forms occur within the passage and chambers of the same monument, mainly within the inner passage and chamber areas. The eastern passage at Knowth measures *c.* 30 m and both the curvilinear and rectilinear art is positioned around the inner passage and chamber areas (*ibid.* 188–195).[11] These images therefore would be near impossible to view from the façade. Further restricted visual access is hampered by a slight kink in the outer passage area. At Newgrange angular and curvilinear art are positioned in the same way and it is probable that an angled light source, especially within each of the passages would cast different shadows depending on the depth of the pecking, the shape of each design and the distance between the image and the torch. Also, angular carvings in the form of infilled lozenges, multiple zigzag lines and triangles dominate the chamber and inner passage areas at Fourknocks. This art too, cannot be viewed from the façade area; the art is hidden and can only be seen from within the central chamber and possibly from the short passage.

A similar design complexity and strategy is found within the British passage grave tradition. At Barclodiad y Gawres (ANG 4) the passage, measuring approximately 6m, leads to a cruciform chamber area which has a series of uprights decorated with chevrons, lozenges, spirals and zigzag designs. These lightly pecked designs are similar to decorated uprights found elsewhere along the Atlantic seaboard. There is also a design association with nearby Bryn Celli Ddu (ANG 1) and monuments found in the Boyne Valley (LYNCH 1967, 1–22). Carved decoration occurs on six stones, located either within the inner passage or chamber areas.[12] Decorated uprights includes Stone 5 (L8), 6 (C1), 7 (C2) 8 (C3), 22 (C13) and 19 (C16) (SHEE-TWOHIG 1981, 229).[13] On Stone 5 are lozenges and vertical zigzags; Stone 6 exhibits a conjoined circular motif, lozenges and vertical zigzags; Stone 7 has a large chevron and on Stones 8 and 19 are a series of

---

[10] These terms in many ways do not capture the style but for the sake of this section of the paper I will use them.

[11] The chamber measures around 20.21 m

[12] POWELL – DANIEL (1956) and later LYNCH (1967) record five stones. However, in March 2006 a team from the University of Bristol discovered a sixth stone that formed the northern upright of the eastern passage. This multi-phased decoration comprised a large chevron that was either superimposed or underlying a series of horizontal pecked lines (NASH *et al.* 2006).

[13] he numbering system uses POWELL – DANIEL (1956), later used by LYNCH (1967) and SHEE-TWOHIG'S numbering (in brackets) (1981).

spirals and a series of unrecognisable motifs. Finally, on Stone 22 is a spiral, with supporting motifs, lozenges, a horizontal chevron band and a series of vertical zigzags. This particular stone bears some resemblance to the decorated *Pattern Stone,* a fallen upright found carved on a fallen upright found over a central pit at Bryn Celli Ddu. Unlike the monuments found in the Boyne Valley, at Barclodiad y Gawres all six stones have their decoration facing into the chamber but, because of the light pecking technique used, the images are hidden from view. A large percentage of the megalithic art within the Boyne Valley appears to be public art, carved onto mound kerbing. Decorated kerbing does not occur though in the Welsh passage grave tradition.

Perceiving Scapes: art, doorways, thresholds and passages

Before I discuss the social (and antisocial) divisions of space within and around the monument, I want to look at the perception given to different spaces and how our senses react in different ways within these spaces. The passage grave builders would have been aware of sensory arousal, of how people using the monument should act and react to certain parts of the monument. The passage grave blueprint, used throughout most of the Atlantic zone during the Middle and late Neolithic, consists of the façade area, the outer passage, the inner passage and finally the chamber. It is clear that as one progresses through these quite different spaces, one's sensory perception changes. In order to move from, say, the façade to the outer passage area, the change of space is represented by both a physical doorway (doorstone) as well as a metaphysical one. Static points such as a stone threshold, constricting or protruding uprights or a lowering of a passage capstone usually represent the physicality of a space. The changes in the physical statementing was arguably accompanied by changes in ambience; changes in light intensity (from light to dark), changes in smell (from open air to stale musty air, due partly to the putrefaction of decomposing flesh) and changes in audibility; from the familiar noises made by the family group within the façade area to the near silence of the inner passage and chamber.

CHRISTOPHER TILLEY's pioneering work on passage grave monuments in Västergötland clearly shows the strategic importance and intentionality of certain architectural features (i.e. what materials are used and how they are used). This group of monuments, consisting of over 265 passage graves within an area of 38 km by 25 km is the most northerly Neolithic group in Europe. According to TILLEY, tombs become 'socialised' through their construction and use, thus allowing sites to become socially-politically manipulated (1991, 68). This process is evident through the changes to the tomb architecture or changes in burial practice. From Tilley's analysis all monuments are standardised in design, comprising an east-west oriented passage leading to a north-south oriented chamber, which are incorporated into a round mound. The mound is delineated by stone kerbing and it is more than likely that the capstone of the chamber were exposed during use. The interplay between the different colours and textures of the stone, the arrangement of the passage and chamber uprights and the way one moves through these different spaces would have been paramount to users of the monument. Not surprisingly many architectural traits are replicated in the two Anglesey monuments and passages along the Atlantic Zone. What is absent though is rock-art.

These single-phased monuments are regularly spaced within the landscape, sometimes in rows of up to twelve and are very visible. TILLEY has also identified a series of intriguing architectural traits that are replicated in most of the Västergötland monuments which further suggests a recognised blueprint in design associated with the ritual use of the monument. Nearly all the uprights used to construct the passage and chamber walls are of sedimentary rock, while the capstones (or roofing stones) are of igneous rock (*ibid.* 70). The entrance to all the passages are narrow, measuring between 0.5 m and 0.8 m in width. The entrance to the passage is also low, suggesting that during use, people entering the monument would have to crawl in order to gain access. However, as one progressed through the passage, the walls and the roof open out and upwards until one reaches the keystone (or threshold). The keystone, located at the transition point between the chamber and passage is a deliberately placed capstone that is lower than the other capstones and, when entering the chamber area, one would have to has to crouch lower in order to gain access to the chamber. Like other thresholds, the keystone is yet another device to restrict visual access from those looking into the passage from the façade area. It also possibly marks the point where the body (or body parts) finally enters the world of the ancestors.

Between the entrance to the passage and the keystone and entrance into the chamber, the body has to travel, albeit a short distance through what TILLEY and others has termed as 'liminal space' (*ibid.* 74–75). This liminal space acts as rite of passage whereby the body is neither of this world or the next. This simplistic hypothesis can be further dissected to represent a series of journeys and not surprisingly adding further complexity to the rite of passage and the way the dead are deposited to the chamber.

Figure 6.1. The façade end of Pentre Ifan. The doorway to the chamber is located between the two supporting uprights (Photograph: GHN).

The door stone at Pentre Ifan in Pembrokeshire, SW Wales stands around 2.2 m above the façade floor level and is located between two uprights which support an enormous capstone estimated to weigh 30 tons (*Figure 6.1*). The doorway, itself weighing around 5 tons is set in such a way that human bone and associated offerings can be placed though the gaps between the door stone and the uprights (CHILDREN – NASH 1997). I have suggested that the door stone may have been physically tilted and moved to one side in order to allow access by certain high status individuals (NASH 2006). Once inside the chamber secret rites could be performed where both the dead and living would have interacted in a special way; what goes behind closed doors! The doorway therefore acts as a thoroughfare to the realm of the supernatural. Special individuals would have the knowledge and power to act as an intermediary between living and the dead. They would deposit the remains of new ancestors, communicate with the old ancestors, replenish the grave goods and offering the new ancestors food and drink for their journey into the next world. However, can the same process of moving the dead across a series of zones be maintained when there are no passages? At Pentre Ifan (Pembrokeshire), the space between the enclosed façade and the chamber is short and the distance between the façade and the chamber, unlike other Neolithic monuments, lacks liminal space. However, one could counter argue and suggest that liminal space is represented by the thickness of the doorstone. This blocking device separates the façade from the large rectangular chamber. Arguably, this journey between the living space and the space for the dead is restricted. Nevertheless, the journey, however short, is still paramount within the act of burial.

Sometimes doorways are set into the fabric of the passage and cannot be moved. Two uprights found within the inner section of the passage at La Hougue Bie, Jersey restricts the visual access between the façade and the chamber (BAAL *et al.* 1925, figure 2). La Hougue Bie, one of Europe's largest and most impressive passage graves and dating to the latter half of the fourth millennium BC has, like other passage graves, have more than one doorway. One of these is located between the passage and the façade while the other is between the inner passage and the chamber. This careful and contrived arrangement of stone uprights within the passage establishes restricted visual access. The builders have added further components to the passage to ensure what is undertaken in the chamber remains secret, visually hidden from people using the façade area. At the eastern end of the passage a deliberate but subtle kink within the length of the passage has been made, and like most passage graves, the passage constricts at the façade end, while at the chamber end the passage walls and roof widen and rise. People using the monument would have therefore been forced to crouch at the entrance in order to gain access to the 9 m long passage. However, as one progressed along the passage, one could move from a crouching position to an upright position by the time one has reached the inner passage and chamber areas. The living and the dead are also guided through the passage and chamber with a set of strategically placed cupmarks (*Figure 6.2*). Several are carved within the inner passage area onto smooth pink granite, while a further 21 are carved onto the eastern face of the northern chamber (*Figures 6.3a & 6.3b*). Arguably, and similar to megalithic art elsewhere these would have been hidden during the Neolithic (assuming that they date from this period). Furthermore, nine subtly-placed cupmarks are located on the roof [capstone] of the northern chamber (MOURANT 1974, figure 3). The strategic location of these cupmarks, between the inner passages to the chamber, possibly demarcates part of the journey for the dead and it is clear that these marks cannot be seen. Indeed, only fire from a hearth or torch could illuminate these marks along with other cultural goods such as decorated pottery and the colour of the sea rhyolite pebble floor (also found within the chamber and passage). Similar to art found at Newgrange and elsewhere in Ireland and Anglesey the cupmarks that are located within and around the northern chamber are hidden from the view of the living and are therefore the property of the dead.

A similar passage arrangement is witnessed, albeit on a much smaller scale at Arthur's Stone, a chambred long mound in west Herefordshire, one of the most notable of all Neolithic burial monuments in western Britain. The site lies on the western intermediate slopes of Merbach Hill and faces the impressive eastern slopes of the Black Mountains of central Wales. This monument, one of the most northerly chambered tombs of the Cotswold-Severn Group, is one of eighteen tombs that dominate the Neolithic landscape of the northern reaches of the Dore, Upper Wye and Usk valleys of Breconshire and neighbouring Herefordshire (CHILDREN – NASH 1994; GRIMES 1936; HEMP 1935). The majority of the monuments within this group conform to a number of architectural rules including locally oriented lateral passages and a false portal located within the façade area.

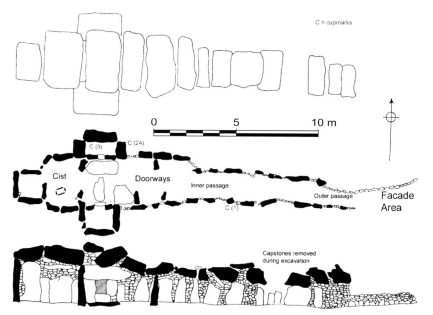

Figure 6.2. Plan of La Hougue Bie, Jersey (adapted from BAAL et al. 1925).

The chamber and passage are set within a cairn rubble mound. The chamber has nine upright stones that support an enormous capstone, estimated to weigh 25 tones. Recent investigations by the author suggest the chamber and passage were set within a long mound that extended some 50 m north of the present mound, probably trapezoidal in shape and similar to other monuments within the group (NASH 2006). At the north-western end of the chamber is an unorthodox

Figure 6.3a. Passage and chamber arrangement illuminated by artificial light (Photograph: Adam Stanford).

Figure 6.3b. Partially hidden upright forming part of the northern chamber showing 21 cupmarks (Photograph: Adam Stanford).

Figure 6.4. The unorthodox passage at Arthur's Stone, Herefordshire.
Passage turns 90° between the façade and the chamber (Photograph: GHN).

right-angled passage with a pronounced doorway, consisting of a single upright (*Figure 6.4*). Between the chamber and a door stone the passage is oriented east. However, from the door stone to the façade and entrance, the passage changes direction to the northwest, pointing towards an impressive scarp of the Black Mountains locally known as Hay Bluff (CHILDREN – NASH 1994, 26; TILLEY 1994, 140). Similar with other monuments mentioned in this chapter Arthur's Stone has a narrow and low entrance that enlarges as one progresses through the passage towards the doorway and chamber. There is clearly no visual access between the chamber and the façade beyond. The 90° kink of the passage alignment where the doorway is sited appears to show a conscious attempt to separate living space and the realm of order and control, from the inside world of death, dismemberment and disorientation. The transition between the two spaces is achieved precisely at that point, equidistant between chamber and entrance, where the passage abruptly changes direction. It is here that the living come into contact with the dead; an act and point in time that is repeated in every chamber monument throughout the Neolithic world. A similar passage arrangement with its restricted visual access is recorded for the majority of passage graves either with or without rock-art.

The three passage graves for discussion, Barclodiad y Gawres, Bryn Celli Ddu and The Calderstones inherently have their problems. The first of these to be considered, and problematic to all monuments of this age, is taphonomy. The surviving material culture probably represents a small percentage of what was within each of the chamber and passage areas. Consumables, such as offerings of food, wood and hide would have long disintegrated over time. This is further hampered by acidic soils that cover most of Wales. Finally, there is a problem with antiquarian and non-scientific excavation methods. All three sites have been severely disturbed; the Calderstones completely destroyed. Thankfully, the two Anglesey sites were excavated during the mid-20th century, albeit after initial antiquarian investigations during the 18th and 19th centuries, and attempts to use the sites as stone quarries (*Appendix 1*). Following excavation both monuments were restored. However, it is not known if the restoration of both monuments was sympathetic with the original layout of the monument. The excavation of Barclodiad y Gawres in 1952–3 was restricted to excavation of the passage and chamber area. Trenching was also extended through the surrounding cairn (POWELL – DANIEL 1956). One can assume that, based on the report, this excavation was by far the most meticulous and scientific. Excavations undertaken at Bryn Celli Ddu, first by Lukis in 1865 and later by Hemp between 1925 and 1929, reveal the probability of a two-phase monument. It is not clear how meticulous both excavations were. What is known is that, based on a mid-19th century engraving, the monument comprised a chamber and passage and the mound had been almost completely removed (used as a quarry). It is not known from the Hemp excavation report if the passage uprights were in their original position or whether or not the outer passage uprights supported capstones. The Calderstones, presently standing out of context in Calderstones Park, has a sadder but well documented history. The site known to have formed part of a parish boundary during the 16th century and is indeed located on a map from this period.[14] Between this period and the 19th century the passage grave were slowly but systematically destroyed. In the early 20th century the remaining stones were placed in storage and later erected within the entrance area of Calderstones Park, before finally being re-erected in a purpose built vestibule in 1964.

Despite problems of taphonomy associated with the early archaeological investigations, one can consider a number of similarities between each of the monuments. Firstly, and most simply, two of the three surviving monuments are architecturally associated with other passage graves within the Irish Sea zone. The architectural style used by monument builders in Anglesey and in Ireland suggests complex contact and exchange networks between these and other groups to the north (COONEY 2000, 226–228; EOGAN 1986, 220). Secondly, are the

---

[14] See Royden on www.btinternet.com/m.royden/mrlhp/local/calders/calders.htm

more complex issues as to the origins of artistic endeavour. The artistic styles from all three monuments are all very different but are nonetheless classified as megalithic art (SHEE-TWOHIG 1981). Each of their styles may have originated from monuments within different areas of Ireland (COONEY 2000).

The artistic style from the Barclodiad y Gawres monument is essentially geometric in form with the predominant designs being chevrons, lozenges and zigzag lines. These occupy four of the six stones that are located within the inner passage and chamber and are essentially hidden from view and cannot be completely seen with natural light. On Stones C3, C13 and C16 large spirals are present while on Stone C3 spirals dominate (*Figure 6.5*). The rock-art from this monument is probably contemporary and despite its unique design coding, many individual design components are also found within Irish passage graves. The meticulous excavation programme undertaken by Powell and Daniel suggests that the position of each of the stones remained *in situ* and what one witnesses today is roughly what was present during the Neolithic.

At Bryn Celli Ddu the art comprises a spiral and, more impressively, a serpentine-style carving (*Figure 6.6*). Both designs configure with Eogan and O'Kelly's curvilinear classification. The small spiral, (13 cm in diameter) located within the southern section of the chamber and carved on an upright may be a later addition. I suggest this because compared with other spirals elsewhere in Anglesey and Ireland, the Bryn Celli Ddu example has been clumsily constructed. The serpentine form, which is carved on what is referred to as the 'Pattern Stone', was discovered lying prostrate over a pit, centrally located and west of the chamber (HEMP 1930). The style of the decoration on the Pattern Stone is similar to Stone 16 at nearby Barclodiad y Gawres and extends over three faces. Excavation by Hemp between 1925 and 1929 revealed an earlier monument phase that he interpreted as being a henge and it is possible and though perhaps doubtful, this stone belongs to this earlier monument. However, I am not entirely convinced that a henge phase exists. Instead the characteristics of the henge may be merely an initial construction or ground preparation phase associated with the passage grave. Whatever the phase, the Pattern Stone and its location, hidden from view, has more to do with the dead rather than the living in that it was found lying prostrate, the art lying face down. Likewise, the spiral, if contemporary with the use of the monument as a passage grave, is difficult to locate and could have only been viewed under certain artificial light conditions.

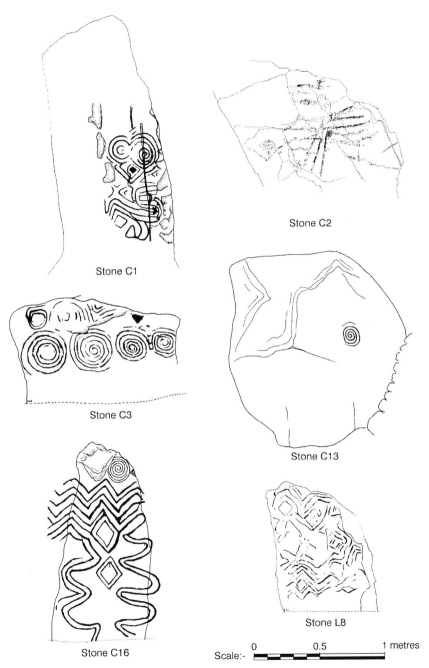

Stone C1

Stone C2

Stone C3

Stone C13

Stone C16

Stone L8

Scale:- 0   0.5   1 metres

Figure 6.5. Decorated stones at Barclodiad y Gawres
(after LYNCH 1967 and NASH et al. 2006).

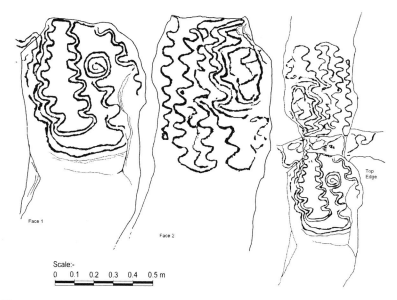

Figure 6.6. Decorated stones at Bryn Celli Ddu (after SHEE-TWOHIG 1981).

Moving through physical and metaphysical spaces

Previously, I have briefly considered the way our senses change when experiencing different types of space (NASH 2006). I now want to focus on visuality and how rock-art may have played a vital role in providing a series of focal points both inside and outside the monument. Of course one can never completely comprehend how and why prehistoric burial was undertaken; there lacks any tangible evidence. A phenomenological approach and the way monuments are approached, has been undertaken by TILLEY (2005), while a more personal account of how people and the dead move through the monument has been effectively discussed by PARKER PEARSON – RICHARDS (1994). Both approaches extend beyond the realms of empirical discourse. Cynically though, the evidence used is limited to just >5% of the potential material culture. However, the architecture of the passage grave tradition is similar throughout the Neolithic core areas of Europe and in each area there is a limited number of sites whose condition fares better than others. From these sites one can ascertain a 'sense of occasion'; the way one approaches the monument and the way one enters the façade, passage and chamber. The ambience for each of these is usually different depending on the condition of the monument. Fortunately, both Barclodiad y Gawres and Bryn Celli Ddu are enclosed within a rubble mound, albeit reconstructed; thus the passage and chamber components are hidden from the outside.

In experiential terms MICHAEL SHANKS has probably come close in attempting to understand ambience and the rhetoric of space (1992). Shanks' retrospective accounts of visiting the Cotswold-Severn monuments of Maes y Felin and Tinkinswood in South Wales and Dunstanburgh Castle in Northumberland is personal and subjective. He encounters though, an incomplete past, a pastiche of preconceived ideas, that affords a casual glimpse of the present. The experience of approaching and entering the monument would have been somewhat different during Neolithic times. One can assume, based on ethnographic and historical evidence, that people using the monument during Neolithic times were high status individuals, and having a similar role to the modern day priest, would have possessed a knowledge and view of the world that would have made him or her different to others. The experience would have been sublime, fulfilling a plethora of human emotions based on terror with fascination, as each space is encountered and experienced. Probably accompanying the priest would have been other individuals of status who would also have special sacred knowledge of the dead. The architectural traits such as the constricting passage the doorway and threshold, and the kink in the passage would have restricted the visuality between the façade and the chamber. Furthermore, the natural light availability would have diminished as one progressed along the passage, into the chamber, adding further visual restrictions on people standing within the façade area. One can see a similar visual restriction within the medieval church. A strategically placed rude screen constructed between the nave and the congregation and the altar area, plus various parts of the act of service being performed by the priest with his back to the congregation, would have established a clear divide between what can and cannot be seen (and understood).

The way the living and the dead move through a *scape* or physical space have implications on the way a space is perceived. Each space, either representing a different architectural space or a different stage in the journey from landscape to the spirit world, creates a different experience or ambience. Both the architecture and the art would have played a vital role in how the dead changed physically and metaphysically as they moved from one space to another.

Concerning burial deposition, it is now becoming clear that the passage grave tradition represents a corporate mentality towards death. However, both Anglesey passage grave monuments show little evidence of this. It is probable that corporate mentality actually means the burial of a high status extended family and that each member would have to undergo a series of processes before he or she could enter the next world. I would stress here that the final resting-place is not the chamber, which merely forms one stage before the metaphysical journey begins.

The journey for the dead begins outside the monument, between the moment of death and entrance to the façade area. It is probable that the monument was outside the main settlement area and would have been approached in a certain way using a series of markers within the landscape. According to the limited radiocarbon dating evidence available the passage grave tradition in Britain and Ireland suggests many monuments were in use between the Late Neolithic and the Early Bronze Age (3000–2000 cal. BC). At the same time other monuments were being used such as standing stones, stone rows and stone circles. In the case of Bryn Celli Ddu two standing stones are located nearby and are possibly contemporary with each other (NASH *et al.* 2006). It is conceivable that the standing stones provided a pathway to Bryn Celli Ddu from the east; I regard this as the first journey. Once at the monument, the body would enter the façade area and thus establish a second journey. Whilst in the façade area, the dead may take on a different guise. There is evidence from other Neolithic burial sites in England and Wales of excarnation platform such as at Gwernvale in Breconshire and Wayland's Smithy in Berkshire. The body may have lain in state over many weeks, at the same time providing a necessary period for the mourning of the deceased. During this time the body would have transformed beyond human recognition, becoming a skeleton, physically leaving the world of the living and entering an ancestral world.

When the time was right the remains would have been collected and transferred from the façade to begin the third journey, along the passage or what TILLEY refers to as liminal space (1991, 75). Liminal space, defined as a metaphysical as well as a physical entity where the dead are neither human nor ancestral, would be divided into two spaces, the inner and outer passage areas. Between the entrance and the inner passage, the dead and its entourage would experience a number of visual and audible sensations. Firstly, the natural or artificial light from the façade area would obscure parts of the passage architecture as one progressed toward the inner passage area. The amount of illumination from this light source would begin to fade. At the same time the noises from the façade would get fainter as the entourage entered into the world of ancestors and spirits. DEVERAUX (2001) suggests that certain stones, either *in situ* architecture or movable, may have been struck that could resonate around the passage and chamber adding further sublime sensations for the occupants of the tomb. These noises, along with chanting both inside and outside the monument, would have created an audible sensation that would have been both synchronous and harmonious. It is at this point within the area of liminal space that the entourage would have to light torches or construct a hearth, probably further within the chamber area. When approaching the inner passage area the rock-art would be fully illuminated. The carved abstract images each with their own meaning would move [and dance] as the flames from the fire flicker. A large hearth area was exposed during the POWELL and DANIEL'S excavation within the chamber area at Barclodiad y Gawres (1956, 16–17). Similar evidence has been found at the highly decorative monument of Gavr'inis, in Brittany. Here, charcoal has been found in both the chamber and passage areas (LE ROUX 1985).

Light emitted from the Barclodiad y Gawres hearth would have illuminated nearly all the stones, including the newly discovered Stone C2 as well as Stone C3, each forming the northern and eastern walls of the eastern chamber, as well Stones C13, C16 and L8. The last two stones flank the doorway between the inner passage and the chamber and can only be seen from these two areas (*Figure 6.7*). The ritual activity that would have been undertaken within the chamber probably remained exclusive to people using this part of the monument. As suggested earlier, using Tilley's analysis on monuments from Västergötland, restricted visual access would have been in place. The carved symbols, their location within the chamber, and their position on the panel, along with their relationship with

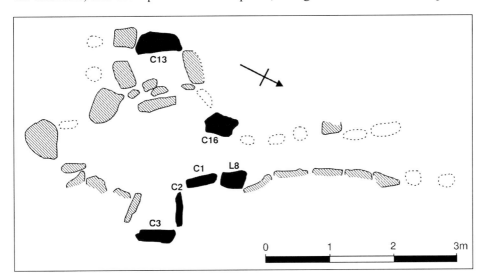

Figure 6.7. Floor plan of the chamber and passage at Barclodiad y Gawres showing the position of the decorated stones (adapted from SHEE-TWOHIG 1981).

other symbols, would have been restricted. In the cases of the two passage graves in Anglesey, and probably the Calderstones monument, the art is positioned in such a way that it could not be seen from the façade or the outer passage. Many passage grave monuments with art appear to conform to this basic rule. It is clear that in some cases, such as the northern chamber or cell of La Hougue Bie or the eastern chamber at Barclodiad y Gawres the rock-art is very difficult to view. It is probably this art was to be viewed only by the ancestors before they or their spirits embarked onto their final journey to the otherworld. The finely pecked lozenge and spirals on Stones C2 and C3 respectively suggest that the light emitted from the hearth would not be enough to completely read and therefore, understand each narrative. It is probable that torches would be required in order to fully illuminate these panels. Once the remains of the dead had been interred, grave goods would be offered in order to accompany the dead on their final journey. Based on the fragmentary evidence from both excavations, offerings would have included pottery vessels, flint tools and, in the case of the burials at Barclodiad y Gawres, a finely carved bone pin (LYNCH 1969b, 158, fig. 21). However, based on ethnographic evidence it is thought that the spirits of the ancestors would not have entirely left the monument. According to many contemporary societies the ancestral spirits could be summoned back from time to time, maybe to assist with later internment. The rock-art, which was hidden, static and probably possessing a restricted meaning would have guided generation after generation of ancestors using a prescribed way of moving through the various spaces, maybe emerging from the art itself. It is clear that many passage graves that possess megalithic art have panels that are entirely hidden from view. At Gavr'inis, Newgrange and in some respects, at Barclodiad y Gawres and Bryn Celli Ddu, certain panels are hidden and inturned towards the monuments' mound. Some of this art is extremely elaborate in form (EOGAN 1986; HEMP 1930; O'KELLY 1982). It could be the case that this inturned art originates from earlier monuments. At the Table des Marchands passage grave, Morbihan, one third of an enormous menhir has been incorporated as a capstone into the chamber architecture. Carved onto the stone is an axe (*Figure 6.8*). The other two sections of the stone have been used capstones that are incorporated into the roofs of Gavr'inis and nearby Er Grah (*Figure 6.9*). However, these sections containing once continuous representative art, including two goats and a large axe, have its art hidden. The destruction of the menhir and its incorporation into a later monument appears to be a deliberate act. Moreover, the positioning of certainly two of the three sections conform with how megalithic art is being used in passage graves.

*Figure 6.8. Central section of an enormous capstone,
later incorporated into the Table des Marchands passage grave (Photograph: GHN).*

*Figure 6.9. The three sections of a menhir,
each section now incorporated into later burial monuments (after LE ROUX 1985).*

Concluding remarks: The Symbolic Role of Fire

In this chapter I have concentrated on the passage grave tradition, focusing mainly on how the living and the dead may have viewed and experienced megalithic art using the available light sources. I have suggested that rock-art, located mainly within the inner passage and chamber areas of two of three passage graves is strategically placed and can be only viewed when using non-natural light sources. The strategic location of art appears to be of paramount importance within the Irish passage grave tradition (O'KELLY 1982; EOGAN 1986). The art can only be viewed and read if the reader is positioned in a particular place inside the monument. I stress, however, that the ambience experienced by people moving from the façade into the outer and inner passage areas and finally, in the chamber area, changes forming part of a special experience that is restricted to only a few. These changes affect all the senses as people move through the different physical spaces, and this is arguably the case whether or not rock-art is present. Rock-art merely forms a series of focal points for the living and, in particular for the dead, as they move through the different spaces whilst on their final journeys to the spirit world. This ambiance is assisted with either strategically placed torch or heath. The latter is present within the archaeological record at Barclodiad y Gawres (POWEL – DANIEL 1956, 16– 18).[15] It would have been paramount that a hearth existed within the central gallery area as no art could have been seen without this light source.

I would further suggest that similar emotions would have been experienced outside the monument as well, and that both the living and the dead were required to undertake a series of journeys that include moving through open spaces via a series of landscape markers, themselves sometimes decorated with rock-art. Hearths appear to be an indicative feature within the façades of many Middle and Late Neolithic burial monuments (*Figure 6.10*). These markers provide rigid points through which people have to move in a particular way. Similar to the way we bury our dead, people are required to follow protocol and funeral rites are properly observed. Markers, usually in the form of standing stones and more subtly, exposed rock-outcropping, would have provided a series of way points whereby the dead (accompanied by the living) made a series of journeys before embarking on their final spiritual journey from the chamber to the next world. When in the chamber area, an array of grave goods would have accompanied the remains.

---

[15] According to the excavation report, over 200 bone of small mammals, fish, reptiles and amphidians were found within this hearth (1956, 16).

Figure 6.10. A reconstruction of a burial ceremony within the façade area of Pentre Ifan, Wales (illustration by Ellie McQueen).

I have suggested that changes in perception are controlled and manipulated by changes in architecture and that how our senses react would depend on the ambience of each space encountered. Likewise, the dead also change both physically and metaphysically as they embark on their journeys through these different spaces, from a fleshed and recognisable corpse, through to decomposition and putrefaction to finally a collection of bleached bone; unrecognisable to those who are involved in the performance of burial.

Incorporated into inner spaces of the monument was rock-art, usually in the form of a series of abstract geometric forms. Although the syntax is limited to, say, ten distinct symbols, the way they are positioned on the panel suggests a complex grammar was in operation. The dominant symbols from the three monuments, including chevrons, concentric circles, cupmarks, lozenges, serpentine forms, spirals and zigzags are located on uprights that are visually restricted and appear to function in areas of the monument where probably only a few individuals would attempt to go. Alternatively, the responsibility of moving the dead from a known to an unknown space would have been restricted to may be one or two high status individuals. As the burial monument is used over many generations the original meaning of each of the symbols may have been forgotten or elaborated on. In order to read each of these symbols the strategic position of a hearth or torch would have been paramount. The flickering flames from a hearth or torch would have been the only means of illumination. This flickering would have made the art move, even dance in front of its restricted audience, transcending its ghost-like images into their physique, until the next visit to this most foreboding place.

Finally, like all architecture, the passage grave tradition indicates that there is an intentionality in design. The design elements repeated throughout the Atlantic zone including mound shape, passage and chamber plans, and the construction methodology would have had a significance to the people using the monument. The idiosyncratic and subtle position of, say, the doorway jambs, the constricting passage, or the location of the rock-art, indicate a status-driven society where by only elite individuals get the privilege of accompanying the dead on part of their journey and experiencing the light at the end of the tunnel!

Acknowledgements

I would like to thank the following people for assistance with this chapter. Firstly sincere thanks to Mihael Budja for inviting me to the 12th Annual Conference at the University of Lublijana, Slovenia where elements of the paper were presented. Also thanks to Abby George and Laurie Waite for taking the time to read through the text. Finally thanks to Dr Becky Rossef for invaluable comments concerning the issues raised in this rather difficult and personal project.

Bibliography

BAAL, A. D. B. – GODFRAY, A. D. B. – NICOLLE, E. T. – RYBOT, N. V. L. 1925
    La Hougue Bie, *Société Jersiaise Annual Bulletin,* 178–236.

BARKER, C. T. 1992
    The Chambered Tombs of South-West Wales: A re-assessment of the Neo-lithic burial monuments of Carmarthenshire and Pembrokeshire. Oxford, Oxbow Monograph 14.

BECKENSALL, S. 1999
    British Prehistoric Rock-art. Tempus.

CHILDREN, G. C. – NASH, G. H. 1994
    Monuments in the Landscape: The Prehistory of Herefordshire, Vol. I, Hereford, Logaston Press.

CHILDREN, G. C. – NASH, G. H. 1997
The Neolithic Sites of Cardiganshire, Carmarthenshire and Pembrokeshire, Vol. V, Hereford, Logaston Press.

CHILDREN, G. C. – NASH, G. H. 2001
Monuments in the Landscape: The Prehistory of Breconshire, Vol. IX, Hereford, Logaston Press.

COONEY, G. 2000
Landscapes of Neolithic Ireland. London: Routledge.

COWELL, R. 1981
The Calderstones: A prehistoric tomb in Liverpool. Merseyside Archaeological Trust.

DANIEL, G. E. 1950
The Prehistoric Chambered Tombs of England and Wales, Cambridge, Cambridge University Press.

DARVILL, T. – WAINWRIGHT, G. 2003
A Cup-marked Stone from Dan-y-garn, Mynachlog-Ddu, Pembrokeshire, and the Prehistoric Rock-art from Wales, *Proceedings of the Prehistoric Society.* 69, 253–264.

DEVERAUX, P. 2001
Stone Age Soundtracks, The Acoustic Archaeology of Ancient Sites, Vega.

DUTTON, A. – CLAPPERTON, K. 2005
Rock art from a Bronze Age cairn at Balblair, near Inverness. Notes in *Past* No. 51 (November).

EOGAN, G. 1986
Knowth and the passage-tombs of Ireland. London: Thames & Hudson.

FORDE-JOHNSON, J. L. 1956
The Calderstone, Liverpool. In: Powell, T. G. E. – Daniel, G. E. (eds.), Barclodiad y Gawres: The excavation of a Megalithic Chambered Tomb in Anglesey, Liverpool: Liverpool University Press.

GRIMES, W. F. 1932
Prehistoric Archaeology in Wales since 1925. The Neolithic Period, *Proceedings of the Prehistoric Society of East Anglia* 7, 85–92.

HEMP, W. J. 1926

The Bachwen Cromlech, *Archaeologia Cambrensis.* (1926), 429–31.

HEMP, W. J. 1930

The Chambered Cairn of Bryn Celli Ddu, *Archaeologia*, 1xxx (1930) 179–214.

HEMP, W. J. 1935

Arthur's Stone, Dorstone, Herefordshire, *Archaeologia Cambrensis.* XC, 288–92.

HEMP, W. J. 1938

Cup Markings at Treflys, Caernarvonshire, *Archaeologia Cambrensis* xciii (1938), 140–1.

HERITY, M. 1970

The early prehistoric period around the Irish Sea. In: Moore, D. (ed.), *The Irish Sea Province in Archaeology and History*. Cambrian Archaeological Association, 29–37.

LE ROUX, C. T. 1985

Gavr'inis: Guides Arechaeologiques de la France. Ministere de la Culture.

LYNCH, F. M. 1967

Barclodiad y Gawres: Comparative Notes on the Decorated Stones, *Archaeologia Cambrensis*, CXVI, 1–22

LYNCH, F. M. 1969a

The Megalithic Tombs of North Wales. In: Powell, T. G. E., Corcoran, J. X. W. P., Lynch, F. – Scott, J. G. (eds.), *Megalithic Enquiries in the West of Britain*, Liverpool, Liverpool University Press, 107–148.

LYNCH, F. M. 1969b

The Contents of Excavated Tombs in North Wales. In: Powell, T. G. E., Corcoran, J. X. W. P., Lynch, F. – Scott, J. G. (eds.), *Megalithic Enquiries in the West of Britain*, Liverpool, Liverpool University Press, 149–174.

LYNCH, F M. 1970

Prehistoric Anglesey, Anglesey Antiquarian Society, Llangefni.

LYNCH, F M. 1972

Portal Dolmens in the Nevern Valley, Pembrokeshire. In: Lynch, F. – Burgess, C. (eds.), *Prehistoric Man in Wales and the West*, Bath, Adams and Dart, 67–84.

MOURANT, A. 1974
Reminiscences of the excavation of La Hougue Bie, *Société Jersiaise Annual Bulletin* 21, 246–53.

NASH, G. H. 2006
The Architecture of Death: The Chambered Monuments of Wales. Logaston Press.

NASH, G. H. – BROOK, C. – GEORGE, A. – HUDSON, D. – MCQUEEN, E. – PARKER, C. – STANFORD, A. – SMITH, A. – SWANN, J. – WAITE, L. 2006
Notes on newly discovered rock-art on and around Neolithic burial chambers in Wales. *Archaeology in Wales.*

O'KELLY, M. J. 1982
Newgrange: Archaeology, Art and Legend. London, Thames & Hudson.

O'SULLIVAN, M. 1986
Approaches to passage tomb art, *Journal of the Royal Society of Antiquaries of Ireland* 116, 68-83.

O'SULLIVAN, M. 1993
Megalithic art in Ireland, Dublin, Town House.

PATTON, M. 1995
New light on Atlantic seaboard passage-grave chronology: radiocarbon dates from La Hougue Bie (Jersey), *Antiquity* Vol. 69, No. 264, 582–586.

PARKER PEARSON, M. – RICHARDS, C. 1994
Architecture and Order: Spatial Representation and Archaeology. In: Parker Pearson, M. – Richards, C. (eds.), *Architecture and Order. Approaches to Social Space*. London & New York, Routledge, 38–72.

PIGGOT, S. 1954
Neolithic Cultures of the British Isles. Cambridge, Cambridge University Press.

POWELL, T. G. E. 1973
Excavation of the Megalithic Chambered Cairn of Dyffryn Ardudwy, Merioneth, Wales, *Archaeologia* 104, 1–49.

POWELL, T. G. E. – DANIEL, G. E. 1956
Barclodiad y Gawres: The Excavation of a Megalithic Chambered Tomb in Anglesey, Liverpool, Liverpool University Press.

RAMOS MUNOZ, J. – GILES PACHECO, F. 1996 (eds.)
El Dolmen de Alberite (Villamartin). Aportaciones a las Formas Económicas y Sociales de las Comunidades Neoliticas en el Noroeste de Cádiz. Cádiz, Universidad de Cádiz.

RCAHM (W) 1997
An Inventory of the Ancient Monuments in Breconshire (Brycheiniog): The Prehistoric and Roman Monuments (Part II). London, HMSO.

SHANKS, M. 1992
Experiencing the Past: On the Character of Archaeology. London, Routledge.

SHARKEY, J. 2004
The Meeting of the Tracks: Rock Art in Ancient Wales. Carreg Gwalch.

SHEE-TWOHIG, E. 1981
Megalithic Art of Western Europe. Oxford, Clarendon Press.

SKINNER, J. REVEREND. 1802
Ten Days Tour through the Island of Anglesey (reprinted with introduction by T. Williams, 2004).

THOMAS, J. 1988
Rethinking the Neolithic. Cambridge, Cambridge University Press.

THOMAS, J. – TILLEY, C. 1993
The axe and the torso: symbolic structures in the Neolithic of Brittany. In: Tilley, C. (ed.), *Interpretative Archaeology*. Oxford, Berg.

TILLEY, C. 1991
Constructing a ritual landscape. In: Jennbert, K., Larsson, L., Petre, R. – Wyszomirska-Werbart, B. (eds.), Regions and Reflections (in honour of Marta Stromberg). *Acta Archaeologica Lundensia* Series 8, No. 20, 67–79.

TILLEY, C. 1994
A Phenomenology of Landscape, London, Berg.

TILLEY, C. 2005
The Materiality of Stone: Explorations in Landscape Phenomenology. Oxford, Berg.

APPENDIX 1: PASSAGE GRAVE DESCRIPTIONS[16]

Barclodiad y Gawres, Llangwyfan

Large passage grave, also referred to as Mynnedd Cnwc or Mynydd y Cnwc, lies 19m AOD on the southern part of a small promontory headland overlooking a small inlet known as Porth Trecastell (SH 3290 7072). This monument appears to have suffered damage until its excavation in 1953. Used as a stone quarry in the 18th century, most of the contents, including archaeological deposits from the chambers, were removed. However, the monument did receive some archaeological recognition in 1799 when a *note* was published by David Thomas in the *Cambrian Register*, listing the Cromlechau or Druidical Altars of Anglesey. One of the first accounts of this monument was given by the REVEREND JOHN SKINNER who on Monday December 6th 1802 described it thus:

> *"Instead of a cromlech at Mynnedd Cnwc we found the vestiges of a large carnedd; many of the flat stones of the cist faen or chamber are still remaining but the small ones have been almost all removed to build a wall close at hand. On another fork of the peninsula about a hundred yards distant we observed the traces of another carnedd of much smaller dimensions [This is now regarded as a Bronze Age cairn]. From the nature of their situation, the bay, the earth work &c. it is not possible to suppose that an engagement here took place with the natives wherein some principal officers were slain and interred on the spot."*

In 1869 H. Pritchard published a full description of the site including a plan of the passage and part of the chamber, but made no reference to the destruction of the monument. The monument was later photographed by J. E. Griffith in 1900. In 1910, E. N. Baynes concluded that Barclodiad y Gawres was a small cairn, and it was not until 1937 that the full extent of the chamber area and the mound was exposed in the plan made by W. F. Grimes. In his research, Grimes concluded that the monument was a passage grave of the style 'of Newgrange and other Irish sites'.

The forecourt area opens out onto views across the western coast of Anglesey. Based on pre-excavation photographic evidence, the chamber and passage architecture prior to excavation was not fully covered. However, when constructed it was probably covered by a large turf mound. The profile of the passage appears to narrow as one progresses into the chamber area, an arrangement unlike passage tombs elsewhere. However, one

---

[16] Although without a passage, nearby Bryn yr Hen Bobl, Llanedwen (SH 5190 6900) possesses many passage grave traits such as a sub-circular mound and enclosed façade. However, no rock-art is present.

should be cautious in so far as the true lines of both passage walls may no longer be in their original positions.

The passage measures approximately 6m and leads to a cruciform chamber which has a series of uprights decorated with chevrons, lozenges, spirals and zigzag designs. These pecked designs are similar to those found within the Boyne Valley and nearby Bryn Celli Ddu (see above).

The cremated remains of two young adult males were found in the western chamber during the 1953 excavation. According to POWELL – DANIEL (1956) no primary pottery was found, but there was one artefact, which may be contemporary with initial use of the monument. This was a bone or antler pin which was found with the cremation burials in the western side chamber, a pin similar to skewer pins found at Loughcrew and Fourknocks in central Ireland. The pin fragments were all burnt and would appear to be associated with the cremation. The location of a cinerary urn, above the collapsed roof area, also suggests that the deposition of the cremation was subsequent to the initial use of the tomb. The urn had a decorated bevelled rim made up of a series of lines of plaited cord impressions. Within the central chamber was discovered a hearth approximately 1m in diameter which contained a mixture of charcoal and stone chips. Also recovered was an assemblage of shells, fish bones, amphibia, reptiles and small mammals.

Bryn Celli Ddu, Llanddaniel-Fab

Bryn Celli Ddu is located on a low ridge of a glacial moraine at around 33 m AOD and close to the Menai Straits and extensive views of the Snowdonia peaks. To the north and west is a slightly undulating landscape. Other Neolithic monuments are sited within the locality including the dolmens of Plas Newydd, Bodowyr and Perthi Duon.

Based on the passage grave sequence in central Ireland Bryn Celli Ddu probably dates from the Late Neolithic and possibly has an association with nearby Early Bronze Age monuments such as the standing stone that is located in a field some 200 m west of the monument (SH 50632 70103). Also worth noting is the recent discovery of 26 cup marks on rock outcropping that lies roughly 250 m north-west of Bryn Celli Ddu (SH 50623 70240).

The mound, 26 m in diameter, may have been possibly larger, but during part-restoration by the Ministry of Works, the monument may have been severely altered. The entrance, with its two uprights (without capstone), is located on the eastern side of the mound. It leads and into a slab-roofed passage approximately 7.5 m in length. Intriguingly, the southern wall of the passage is straight, whereas the northern wall is not (THOMAS 1988, 45). To the west of the monument and almost in line with the alignment of the walls of the passage is a standing stone, suggesting fore planning of the monument. The passage leads into a polygonal chamber roughly 2.5 m across.[17] Between the entrance and the passage are two sets of kerbing, which suggests two phases of passage grave building.

The Reverend John Skinner visited the site in December 1802 and writes an important account of how he entered the passage. The site was excavated by Captain F. du Bois Lukis in 1865 and later by Hemp between 1925 and 1929. Hemp revealed a possible complex multi-phased history to the site. Beneath the mound was, according to Hemp a circular henge consisting of 14 upright stones, some of which were broken others, leaning outwards, and within the centre of this was a pit that was a recumbent stone slab. During excavation, socket holes were found which might represent the position of further uprights. Underlying some of these socket holes was evidence of cremation material. Covering the area of this possible henge monument, but underlying the present mound, was a purple coloured clay that Hemp suggests may represent a ritual floor. However, LYNCH (1969a, 112) argues that it is a palaeo-turf line, the colour of which has been affected by drainage conditions from the overlying mound. Lying next to the pit was the Pattern Stone.

Finds from the 1925–29 excavation are meagre and included a petit tranchet arrowhead that is probably Late Neolithic in date. Also recovered was a rounded scraper (thumb-shaped end scraper), a small lithic assemblage numbering twenty pieces and a mudstone bead which was found within the turf line of the ditch, south of the passage. Previous antiquarian interest in this monument (dating to at least the early 19th century) has probably seen the removal of much of the artefactual evidence from this site.

Within the entrance area of the passage-grave were the sockets of five post-holes that may represent a possible burial platform for human excarnation and two hearths. Immediately behind this structure was a shallow pit containing the remains of an ox burial. The presence of the post-holes, the burial pit and the two hearths suggests ritual activity is an ongoing process whilst the monument was in use.

Along with Barclodiad y Gawres, this monument has two stones with megalithic art, one within a pit, the other in the chamber. Decoration of one of two stones includes an anti-clockwise spiral approximately 13 cm in diameter. The other stone, known usually as the 'Pattern Stone' was found in a possible ritual pit in the centre of the monument. The Pattern Stone was believed by Hemp to belong to the henge phase.

Considering the design and the presence of an ox burial LYNCH suggests that this monument has a greater association with passage graves in Brittany than with monuments in the Boyne Valley in southern Ireland (1969a, 111). Within the chamber area is a single

---

[17] LYNCH (1969a, 116) refers to the chamber as roughly polygonal.

pillar-stone, which has no structural use and therefore may be considered as possessing some ritual or at least, an aesthetic significance.

The complex decoration of the 'Pattern Stone', located on three of its faces comprises a clockwise spiral which is linked to a meandering curvilinear pattern, referred to by SHEE-TWOHIG as a serpentine form (1981, 230). This design covers both faces of the stone. Also present is a cup-mark. The simple spiral may have direct similarities with stone C16 within the chamber of Barclodiad y Gawres and Stones A, B, C, D, E & F of the Calderstones, Liverpool.

Outside the upright stones were the remains of a silted, flat-bottomed ditch approximately 6m in diameter and 2.2 m in depth. It is this feature that is considered to be the henge. However, it is just as plausible to suggest that the henge uprights may in fact represent the kerbing of the passage grave.

Calderstones, Liverpool

This now destroyed site has, in my opinion, the most complex decorated carved megalithic art assemblage in southern Britain. The history of its destruction can be traced as far back as the 18th century. Surviving today are just six decorated stones that once formed the uprights of a passage and/or chamber. Documentary evidence for this site is good and several early 19th century engravings exist of the site which by then was much denuded. The original site is not known and today the stones stand within a glasshouse in Calderstones Park. There have been attempts, albeit based on limited evidence to place each of the stones into their original context (COWELL 1981). Based on the location of decorated stones at Barclodiad y Gawres, Bryn Celli Ddu and monuments within the Boyne Valley in Ireland, the Calderstones' uprights would have been erected within the chamber and inner passage areas.

The stones were first recorded properly in 1954 by FORDE-JOHNSON who labelled the stones A to F (*Figure 6.11*). However, Forde-Johnston and later COWELL (1981) were concerned with recording only the megalithic art. The rock-art present on these stones is divided into at least four chronological phases that include: Phase I – megalithic art includes concentric circles, grooves, lines and spirals. Phase II include Late Neolithic/ Early Bronze Age arcs, an axe, cupmarks, footprints and a wheel motif. Phase III includes medieval and post-medieval graffiti (a Maltese cross, shoeprints and text) and Phase IV includes 20th and 21st century textual graffiti.

Interestingly, the art from Phase II does not superimpose Phase I art suggesting some degree of respect for the earlier tradition. It is probable that both phases are within a generation or two of each other. Many components used within both phases are found

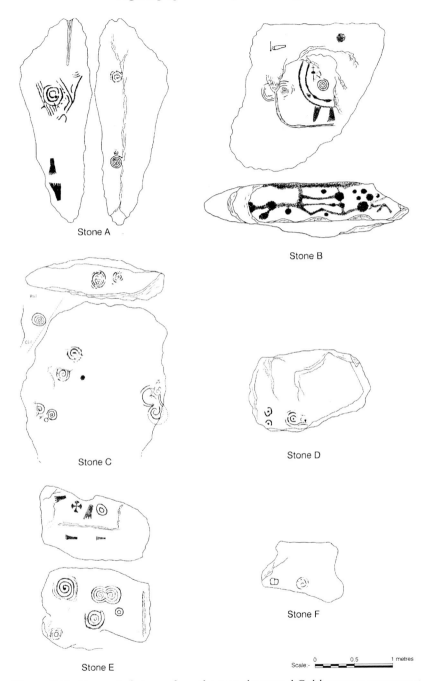

*Figure 6.11. Decorated stones from the now destroyed Calderstones monument
(after SHEE-TWOHIG 1981).*

elsewhere on Late Neolithic/Early Bronze Age exposed rock-art in Northern Britain (see BECKENSALL 1999). Common artistic themes found on all three monuments are spirals and concentric circles and these designs are probably the earliest. Footprints, of which there are four on the Calderstones are considered rare in Britain and are restricted to a limited number of sites. The Calderstone footprints probably belong to Phase II, dating to between c. 2,500 and 1,800 BC. Similar carvings have been found on a capstone belonging to the Pool Farm cist burial in Somerset and this site is clearly Bronze Age in date.

7

Semiological Discourse on the Process of Reduction of Copper Ore in the Copper and Bronze Ages

MARIE-CHANTAL FRÈRE-SAUTOT

Introduction

The transformation of minerals into useable materials, copper for our purposes, follow a rigid procedure which is based today on laws of chemistry, principally mineral, as well as modern physical sciences. Under these conditions, one can imagine that for the "first metallurgists' during the 5th or 4th millennium BC, this indispensable operation for the making of tools, weapons and jewellery had to have been founded on other references. Whether the discovery of metallurgy was linked to the mastery of fire accidentally by a potter bivouacking near a source of minerals, or by the chance loss of a pearl of malachite or azurite in an abode, or through the experiments of a sorcerer's apprentice, all hypotheses are possible in imagining a trick Nature can play on man if not a haphazard meeting two elements, namely: ore and fire. However, in between the spontaneous beginning and the intentional production, there is a long road…that of mastering technique – in other words, the elaboration of concepts and specific schemes of operation. One is immediately tempted to think that the accomplishment of a series of transformative operations could only seem like magic or witchcraft to those men; especially in considering the ancient myths related to the art of the metal worker, the outcome of metallurgy, if not to the ostracism of these workers by certain ancient or traditional societies. Sacred technique or technique of the sacred – one is reminded that medieval alchemy offers a more recent, though notable, facet of the notion: that metallurgy still has a connotation not only of the irrational hidden in all its technical innovations, but also of something supernatural which is the utmost paradox.

The appearance of the first metallurgist, as all archaeologists have more or less remarked, corresponds to two phenomena observed as material testimonies of culture: splendorously rich sepulchres such as the unsurpassable tomb of Varna, which are interpreted as undeniable traces of a great social hierarchy based on a concentration of wealth and the development of megalithism, or of a certain monumentality like the pyramids. From there to conclude that the mastery of metallurgy leads to a change in behaviour, beliefs and social order, as compared to the preceding Neolithic societies, as cautious interpretations of archaeological discoveries reveal, there is only one step for societies still deprived of a tradition of writing.

Figure 7.1. Angled tuyere and crucible. Archéodrome de Bourgogne
(Photograph Dubouloz).

Pyrotechnical science is often considered the essential vector of metallurgy, and Haudricout claims that bellows made of hide fathered the evolution of this domain. When one understands the importance of natural ventilation recently experimented for research, this affirmation is puzzling. Certainly, such a fundamental aspect as the rise in temperature can also be caused by predominant winds. In fact, numerous regions where metal works seem to develop happen to be desert lands swept by strong aeolian winds: Negev, Sinai, the Lut desert in Iran, etc. Other operations are however indispensable to the development of iron making technique, firstly the location and exploitation of an appropriate vein of minerals. Several hundred thousand years of searching for this primary matter undoubtedly helped in developing a sense of observation of mineral sources since prehistoric

Figures 7.2 & 7.3. Experimental oven in Kargali, Russia (Photograph APAB).

*Figures 7.4 & 7.5. Two phases of cooper smelting in a pit as an oven,
Archéodrome de Bourgogne (Photograph APAB).*

men could only rely on oral tradition to acquire any mastery of it. For centuries, this quest surely led to techniques for location minerals underground with the help of wells and galleries. Following a vein of minerals seems as logical as extracting matter from a layer of clay. Carbonated ore is recognizable by its colour: azurite blue and malachite blue green – which are among the most visible in nature. What follows is the essential joining of primary matter with heat, the channelling of heat and finally the control of its temperature. A reductive flame changes colour and shimmers differently according to chemical operations conducted in the hearth and clearly indicates each step accomplished. They observed the dark matter that spewed out of the fire and covered pure metal balls in the black porous magma. Separation and reassemblage constitute two final stages for the obtention of raw copper that is then transformed, occasionally by hammering, but most often by fusion into useful objects (*Figures 7.4 & 7.5*). If the European Chacolithic Age is characterized by the notable presence of copper pearls, it is undoubtedly due to the fact that this most precious material was reserved for jewellery which held greater value than tools or weapons. These latter are second place to the blasé of the dagger, also one of oldest fossil directors of this new 'technical" age. In comparison to stone which is malleable, to weed or bone which breaks, copper is not only ductile and elastic, but it marries the shape of a mould, meaning that it can be reproduced in multiples; also it reflects light – attributes that continuously renew our approach to this material. It is a fact that authors who treat the first Bronze Age or the Copper Age, tend to insist on the symbolic aspect of wearing metal objects which represent possessions in terms of either things or power, and the noticeable consequences these have on the development of agriculture, the dimension of the territory, the foreboding network of long distance exchanges. In fact, all temporo-spatial points of reference are modified by the intrusion of this technology.

It is no accident that Old Testament's recounting of the *burning bush* shows the presence of God as a flame: fire uncovers the profound nature of things. In Greek mythology, Prometheus steals fire from the Gods who have made man out of clay, and it is fire that allows humanity to survive. Also, the fire that burns on the altar or burns the sacrificial offering, serves as the symbol of the relation between man and the Gods. What can be said about the ritual of incineration that came before metallurgy, for which the abstract symbolism replaces the burial of a dead body… Because fire devours and leaves only ashes, the process of metallurgy is one such case, like pottery, where fire produces and transforms. As if fire and matter were indissociably associated, BACHELARD (1938) writes in

The Psychoanalysis of Fire: "To obtain the essence of fire, man must return to his source, to his reserve where he concentrates himself, meaning in the mineral".

As for proper metallurgical operations, where copper is concerned most of the traditions were revisited when iron replaced bronze and proved to be better adapted for the production of sturdier objects, because it is derived from more common primary matter than those contained in copper, and also less expensive than the alloy for producing bronze. Furthermore, even their vestiges attest to the differences between these two metalworks: the little makeshift furnaces employed to reduce copper did not endure while the multiple of those destined for iron production left significant traces whose recent discoveries have provided with new data.

In sacred rituals, fire has always been associated with purification. The elimination of past impurities through a fire ritual in many religions corresponds to the process of metallurgy in a similar way that fire is used to separate the ore from other components. It is important to situate the emergence of these rituals in relation to the techniques in order to determine whether they coincide or which came before the other. Does the relation to the sacred actually lead to the technique? If it is true, as Pétrequin and Fluzin affirm in their introduction to the 1999 Antibes Colloquium (PÉTREQUIN *et al.* 2000), that the first products derived from the arts of fire were dedicated to prestigious uses associated to ceremonies for the living, as well as for the funeral domain, then this only underlines the close relationship of sacred and one can well imagine the comparisons between the processes and their signification, or between the technical operations and the rituals. Perhaps familiarity with these technical systems led to understanding the most essential stages of operations of transformation of the activity of metalworks. As such, experimental attempts might offer us hints of reasoning. To summarize the method without relying on chemistry, the principle operations of metallurgy entail: extract the ore, set up the hearth: dig a simple ditch lined with clay to define a circular oven, break up the ore, place the ore and a combustible (wood charcoal) in the oven, light and ventilate the oven until desired heat is obtained, extract the ore, separate the copper from the residue, re-assemblage and fusion of copper.

In visual terms, we can describe the different phases for the elaboration of metal through variable colouring and the state of the matter in fusion: for example, in carbonated ore, malachite or azurite due to their blue or green colour, even today, attest to the presence of copper oxide. This vivid colour, at the heart of a black mass, changes into brilliant smooth yellow fragments. Re-melted, these fragments turn into liquid, and then in contact with air, they solidify into a form which becomes an ingot (*Figure 7.6*). It seems reasonable to say that through successive actions of fire at least three successive states of matter result with marked changes in appearance. What happens at the centre of a hearth in a restricted space, in the "secret" of a furnace, is like the ripening of a young fruit, solid inert rock turns into a fluid and then resumes its solid state. We know of no other visible transformation in Nature that corresponds to these phases. Yet those prehistoric men who might have witnessed the activity of a volcano perhaps saw a similar effect of fire on matter – their fear of an eruption is another reason to consider the operation of metallurgy as something supernatural.

Figure 7.6. The formation of the ingot, Archéodrome de Bourgogne (Photograph APAB).

However the metallurgist is one who knows, controls and reproduces: in other words, one who masters the process, and therefore in his eyes, even if the logic of our scientific concepts is foreign to him, has the power to act. Carrying this even further, archaeometric analyses show that, even in ancient time, the types of copper or alloys were reproduced and selected in relation to the nature of produced objects so that the degree of hardness could vary just as the colour of fusibility. Even if the potter, who preceded the metalworker, was able to harden clay and render this mouldable material impermeable, it took only one stage to transform his matter from one state to another without entering into a sequence of operations. For pottery, only one heating episode was necessary. Ceramics are like cooking, whereas heating metal requires a different state of abstraction that implies a full understanding of the phenomenon. In other words, if the prehistoric metallurgist did not employ the same concepts as his 21st century homologue, at least he knew exactly what to expect at each phase of the transformation and could intervene to modify or correct the course of actions according to his satisfaction. He chose his ore and combustible in relation to his environment, he determined the shape of his furnace, varied the size of his pounded bits of ore, and varied the quantities and type of ventilation the furnace would use, or opted for natural ventilation. All this signifies that he possessed knowledge of precise effects of each parameter intervening in this operation, knowledge equal to technical know-how. However, it did not suffice to demonstrate it, he had to be able to transmit it by describing each phase including those taking place inside the hearth, out of sight. Any stone cutter can easily dispense with commentary to transmit his knowledge by means of a simple gesture; however the metallurgist has to communicate a highly complex process.

Keep in mind the era when metallurgy appeared; certainly, as has been already pointed out, man was familiar with fire for making pottery, but the incineration of the dead was also a customary practice among certain populations. It must have taken a special power of authority for a man to agree to burning the body of a parent: the destruction of matter serving as a necessary passage of the living into another phase. This symbolic function of fire is probably not familiar to all populations of metalworkers; however, the abstract meaning of the apparent destruction of a material toward another state, like the transformation of water into gas through heat, was confirmed know-how. Familiarity in any case does not diminish the magical aspect of the process. If we adhere to any manipulation through our own experience, experimental palaeo-metallurgy appears more related to science fiction. So to stay within reasonable limits, we shall rather adhere to suppositions concerning only the level of technology that we know for certain about these populations. It would be terribly risky to artificially construct a system of symbolic references based on technical process, although we do have the right to try to understand the level of abstraction which the transformation of ore into metal must have represented, and then to conclude with a general symbolic structure without establishing a cosmogony.

If the emergence of the technique for the reduction of copper ore manifested itself in different places at different periods, nothing has proven that such transmission of know-how between the various hearths ever crossed the Mediterranean. Seeing that the most ancient vestiges come from Turkey, then from the Balkans regions and the Middle Orient, corresponding today to the region from Jordan to the Negev with an equally important centre in Oman, and probably also East of Iran, all within a time frame of one to two thousand years, it s highly plausible that these centres cropped up independently of one another. In other words, the beginnings of metallurgy were inscribed in different cultures, all essentially Neolithic, but with different cultural traditions and various representational systems. For instance, the impact of technical innovation is not the same in a hierarchical society as in one that is not, or in a society that possesses a pyramidal system of beliefs and one that does not, etc. If we attempt to summarize what characterises in expressions of material culture as we know them today, other novelties chronologically corresponding to the same period, we can assume three things: the intensification of urbanism and long distance exchanges, the emergence of grandiose funerary monuments, the huge differentiation among individuals in their funerary context. Yet, this is far from pointing out a relation of causality. Numerous authors insist on the strong representativity of the cemetery of Varna as if its metal corresponded to a new concentration of power, a new hierarchy and easily transportable wealth, differing from land and cattle that constitute the fundamental riches of Neolithic society. It should also be reminded that metallurgy, unlike other techniques, does not dispose of mineral resources which are evenly located over the entire planet. Therefore, by all logic, it is the proximity to a vein of minerals that will determine the appearance of the technical system. It seems though, that among all the regions where metallurgy emerged there is an equitable mastering of the early stages of technique, yet those sectors more inclined to have mined ore are limited to a few pockets dispersed over thousands of kilometres. This is why it can be assumed with no other proof, that metallurgical societies, although they might have kept their technical process confidential if not secret, were all open because commerce and exchange presided in this type of work. On one hand, the ore or ingots had to be transported, and on the other hand, the products were commercialised within a certain radius, however large. If we care to consider the social implications of metallurgy on an ethic group and the symbolic implications it holds, we can make comparisons with studies that have been published on African metalworking societies. If ethnoarchaeology does not offer an answer to these questions, at least it does offer precious insight.

At first, an analysis of this social distinction which could likely concern certain prehistoric societies could seem trite, however with at closer look it seems that the double speciality of "farmer-ironworker" covers the same reality: man's relation to the earth which is his means of survival. The earth nourishes with plentiful foodstuffs and is no less generous with mineral resources that the farmer transforms into metal just as his seeds sprout after he sows them. The articulation of the Neolithic social structures is only facilitated in this way since the technical mutation does not include a radical social change. The metalworker is neither a wise man nor a magician… in fact he is probably no more than a simple farmer. This distances us from the interpretation that metallurgy effected a change on social relations by establishing a very pronounced hierarchy, yet it is merely hypothetical.

In looking once again at the technical constraints of this process, it would be interesting to explore other metallurgical processes for the production of copper such as sulphide metallurgy. Here we are confronted with a much more complex series of operations as those of carbonates. The reduction of sulphides entails a sequence of phases without which the obtention of a metal would be impossible. This calls for the elimination of sulphur, then the separation of iron and copper through a chain of oxidations and reductions. The first stage of this work consists of one or more grilling procedures that serve to eliminate impurities before the actual reduction that takes place in the reduction furnace.

We will probably never fully understand how the passage of one type of metallurgy to another occurred, though we know that a vein normally carries sulphide ore as well as carbonated elements and that its exploitation necessarily leads to the second variety after exhausting the first. Yet nothing keeps us from supposing, as some have already demonstrated (GMPCA colloquium 2001 – Bourgarit, Mile) that the various metallurgies coexisted from the start and that only the processes developed in a consecutive manner. To arrive to such mastery, copper workers were undoubtedly highly specialized individuals. This is confirmed by the displays of bronze and gold smiths found in certain tombs of Bronze Age like the Genelard Tomb (Saone-et-Loire) for which considerable masses of metal were produced and circulated by a specialized group. The importance of this primary matter and its expansion supports the idea that the social standing of the person who knew how to extract and work metal was not necessarily that of the prestigious person entombed. Therefore, was the peasant, like the African metalworker, a simple artisan or a scientist initiated to secrets? It is tempting to say that at this time there was the birth of heroic tales and founding myths, and that know-how transmitted to man by Gods or by heroes allowed for the establishment of some sort of cosmogony. But it would be negligent to state that the great myths that we have inherited today are rooted in ancient mythologies that existed before written history. Nevertheless, the little that we can interpret from the bits of representation handed down appears to be directly linked to the great cycles of Mediterranean legends and religions.

All arguments around the period of diffusion of the first metallurgy coincide in underlining a significant change in society as well as in numerous founding elements of a new model of social organization and of representations of the world than have endured until today. The punctual localisation of principal mining points undoubtedly contributed largely to the intensification of exchanges, therefore to the diffusion of culture on important realms of action. In questioning the simple postulate of transformation induced in a universe organized by the effects of reduced ironwork, we can suppose that the intrusion of a technical discovery had a direct influence on their way of understanding the world, on social structuration and also beliefs. It is useless to state that fire has an important role in numerous religions. But where metal is concerned, one quickly realized that it carries a certain number of symbols inscribed in different myths in which the blacksmith's role or (and) forged utensils are found.

Most certainly, in the realm of know how, he who possesses such a learned technique as metallurgy, holds a particular status at the centre of a group either because science inspires respect or else uncommon knowledge is capable of stirring such powerful fear that the person in question is excluded from the community. If we conceive this person as an itinerant metallurgist taking his own know-how from one place to another, then we can well imagine this traveller being met with suspicion, since a foreigner always sparks mistrust especially if he possesses knowledge and secrets. In this case his status joins that of the sorcerer, magician or shaman... though these are only suppositions.

Important points in our archaeological understanding: in large western societies where the first characteristic copper tools appeared, for instance the axes of Remedello, the pearls and awls of Chalcolithic Mediterranean, the arrowheads of Palmella diffused widely throughout the Mediterranean, it seems that profound changes affected ceramics, habitats, funerary rituals and representations..., yet the eventual link with this technical innovation is not known. Also, the expansion of a well-characterized "culture" coinciding to the Chalcolithic must be considered that is the European bell shape (*campaniforme*) which gave rise to numerous interpretations and which showed great originality. The rupture that resulted with the installation of these groups in the Alpin milieu is particularly evident in the case of the site of the Petit-Chasseur (Little Hunter) in Sion with the work of Alain Galley who often used data taken from ethnography to interpret the bell-shape phenomenon. On this topic, he insists quite a bit on the close relation between the hierarchical structure of a society set up as a chiefdom and the exemplary megalithic representations it has left behind. In view of a persistent Neolithic economy founded on agro-pastoral resources, megalithism might well represent a new economic challenge if not a strategy destined to reinforce the power of a few. All ethnographic examples brought up by GALLEY (1995), like those quoted by OTTAWAY (1994), confirm this link between the metallurgical society and the concentration of power. As for the role of the metalworker, examples drawn from African ethnic groups highlight the special place the ironworker has in society and the roles played by the creative forces within his social stature that are accompanied by prohibitions. Of course, what fundamentally distinguishes the metallurgist from other citizens is the extreme specialisation and inaccessibility to his art by other beings in the same community. Although stone cutting is also a specialisation, all those who carried weapons or tools were undoubtedly capable of sharpening their edges themselves. In this way technical knowledge of the material at hand allowed him to conduct an act of production. One who possesses a copper axe can easily sharpen the edge but this does not constitute an act of production. The strong metaphoric value of the first objects made of copper overshadows their technical function and their commercial value. This is not necessarily the case of these first productions that might have undergone an intense system of exchange: certainly for the attribution of power and signs of prestige, marketing value pending. Without a doubt, the metallurgist had an intermediary role between Gods and mortal even if his production was part of a banal system of objects. The Chalcolithic dagger of the steles of Ligure or Trentino has a great symbolic significance; this is also seen in the anthropomorphic steles in the great cycles of engraving of Mont Bego or in the Valley of Wonders. It is a complex technical object and an effective weapon. However, fire that is basic to its production, is totally absent in these representations, as if the object itself, like for all of them, was solely invested as the symbol of fire.

We must put the metallurgical operation aside in order to try to examine the multiple facets of the contribution of the first metallurgies and the semiological upheavals which can be deduced from archaeological vestiges. Fire that heats, lights, cooks and protects prehistoric societies suddenly produces a primary matter from which fundamental objects are conceived for man and for display: weapons, tools, jewellery. Here it is invested in a complementary function directly joining the cremation fire that destroys the body in order to let the dead person pass into another realm of reality. Systems of beliefs in the "proto-scientific" relate to cults.

Difficulties involving study of these intermediary periods and the fleeting aspect of copper work operations explain why rarely in European archaeological literature the problem of metallurgical fire and significations is no longer treated. It must be pointed out that we are working with particularly unclear data wherein both technical and social aspects are concerned. Certainly some signs are proof of significant transformations such as funerary usages, coherent groupings like the bell-shaped "set", important changes in pottery with the arrival of a "European concept", and the transmission of all these traditions throughout the Bronze Age, all confirm the great influence of the Chacolithic societies.

Nevertheless, it is not totally risky to suppose that the metallurgists' fire transformed the relation between man and his comprehension of Nature by reinforcing his own individual power, in showing his capability to transform something far beyond its appearance. Perhaps this is reason that the Greeks' myth presents Prometheus as the one who steals fire from the Gods.

If there is one book that takes into account the close link between the ore and metallurgy, it is "Blacksmiths and alchemists" by MIRCEA ELIADE (1977) who makes an inventory of the myths and legends linked to mining, extraction, transformation of ore into metal into weapons, etc... He establishes a constant relation between metal technique and the sacred meaning of the metallurgical operations throughout the ages and cultures, and an attempt to establish a line between metal work and alchemical research. Fire – divined, sexualised, endowed with great symbolic value, reappears many times in his book, but it is in the following paragraph that he precisely defines its semantic value: "It is above all with fire that we change Nature, and it is significant that the mastery of fire affirms itself in the advance of tributary cultures just as in psychological techniques which are basic to the oldest known magicians and shaman mystics." Existing for over twenty-five years, this reference to the shaman is still applicable today since it joins prehistoric artistic creations to behaviours stemming from shamanism. Naturally research on esotericism, the hidden or secretly revealed, invites another observation: that metallurgy started and developed in areas bordering the Mediterranean within the great chains of Taurus, Zagros and the Carpathians, the Arabian shield, in a zone which has proved to be the privileged domain of religions of mystery, the famous oriental cults from Zoroastrism to the cult of Mithra, from which are more or less derived the great monotheisms of the last three millenniums all holding a common base of imagery and tales. Can it be said that it is not by accident, that technical knowledge and spiritual revelation can be coincidental? We are entirely unable to confirm these reflections; they are rather interpretations that distance us from ordinary material preoccupations. In archaeology we can only rely on material facts, fragments most often void of accompanying texts at least for the periods that interest us: material for thought is absent. To avoid transforming our vision of ancient societies into an inventory of dispersed objects, it is sometimes necessary to rely on a quest for the meaning that the interpretation of symbols might offer. Probably we will never find sufficient elements of proof; for this, metallurgy would have to be directly associated with visible elements of a belief. The presence of copper objects in tombs like the famous triangular dagger (campaniform) only reveal part of the question, just as pottery or stone jewellery or shells, all of which hold a sacrificial value of the deceased. Although in this case the object is alienated from its production phase and the carrier of the dagger represents nothing in relation to the power of metallurgist. Certainly some tombs contain groupings of which appear to resemble utensils of a metallurgist. As for the enormous mass of deposits, hardly any of them correspond to the very ancient periods, and most of them are far from relating to civilisation. One assemblage of objects, remarkable for its weight, for the quantity of objects it contains, its late dating and technical prowess is that of the Nahal Mishmach (Israel) housed in the museum of Jerusalem. Technical studies show that it is of a homogeneous metal undoubtedly from local mines of the Negev, which were exploited in these mines. Made of a compounded alloy in the 5th millennium BC, the eighty-nine pieces processed by melted wax assembled and tucked into matting were found near a small sanctuary with no clear link between this depository and the place of cult. A great amount of curiosity has raised over this "treasure", or "metal caster's hiding place" – metal depositories to which many meanings have been attributed from a simple treasure to an offering made to the Gods of Nature. But these interpretations are constantly enriched by new discoveries of our patrimony. Precocious technology anchored in an entirely Neolithic environment where the wealth of the earth sufficed to assure the survival of these communities, it is in this environment that man became learned enough to produce copper, and then bronze. When a farmer sows a seed or assists in the birth of a lamb, the produce remains exterior to himself, whereas when he collects a particular stone, brings it back to enclose it in a hearth, to heat it in order to make a new material, he is the master of his creation from one end to the other.

Perhaps this is what permitted him to measure himself to the Gods, or to consider the Gods in the image of man.

Bibliography

BACHELARD, G. 1938
    La psychologie du feu, Paris, Gallimard.

ELIADE, M. 1977
    Forgerons et alchimistes, Paris, Flammarion.

GALLAY, A. 1995 (ed.)
    Dans les Alpes à l'aube du métal, Archéologie et Bande dessinée, Musées cantonaux du Valais.

OTTAWAY, B. S. 1994
    Prähistorische Archäometallurgie, Verlag Marie L. Leidorf.

PÉTREQUIN P. – FLUZIN, P. – THIRIOT, J. – BENOIT. P. 2000 (eds.)
    Arts du feu et productions artisanales, XXe rencontres internationales d'Archéologie et d'Histoire d'Antibes, Editions APDCA.

8

Uses of Fire at the Magdalenian Site of Verberie, France

FRANÇOISE AUDOUZE

Introduction

For this paper, we are interested in the materiality, social and symbolic functions of fire in late Upper Palaeolithic settlements as exemplified by the late Magdalenian site of Verberie (le Buisson Campin) in Northern France, and other Magdalenian campsites of the Paris basin (*Figure 8.1*).[1] These sites are open-air sites repeatedly covered with floods that brought silt that gently covered the occupation layers, letting artefacts in situ. Their spatial organisation is intact with features such as hearths, activity areas, flint workshops, butchering activity areas and dumps (OLIVE *et al.* 2000). The palimpsest effect is minimal due to the short duration of these occupations (*Figure 8.2*). The Verberie site is located in the valley floor of the Oise River, just south of Compiègne. It is made with eight superimposed occupation layers. Every one represents a fall hunting camp of short duration related to the fall reindeer migration (AUDOUZE 1987; ENLOE 1996; ENLOE – DAVID 1995). The archaeological layers are thin occupation floors 3 to 5 cm thick embedded in silt. They have been rapidly covered by periodical gentle flooding which ensured the excellent preservation of bones, lithics and of the spatial organisation. The same type of alluviation process preserved Pincevent and Etiolles sites in the Seine valley (OLIVE *et al.* 2000; VALENTIN – PIGEOT 2000).

Fire is an essential component of the survival and of everyday life of Palaeolithic hunters. It is used for procuring light, heat, for transforming raw materials, whether organic or mineral, vegetal or animal, for cooking food, smoking hides, firing ochre and many other processing that we may not know of. It is present in Upper Palaeolithic settlements as hearths, heated stones, heated

---

[1] The group of sites called the Magdalenian group of the Paris basin includes Picevent (Seine-et-Marne), Etiolles (Essonne), Marsangy (Yonne) and Verberie (Oise) and 12 other sites without faunal remains or spatial patterning. The group will be refered in the following pages as MBP (Magdalénien du Bassin parisien).

or burnt artefacts, ashes refuse areas; the three last categories are refuse from hearths. It also is present in caves as hearths, lamps or torches.

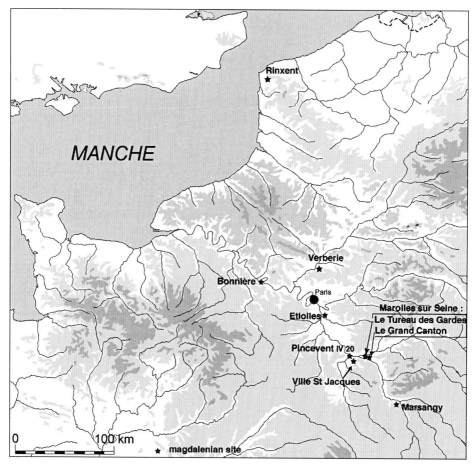

*Figure 8.1. Centre of the Paris Basin with the late Magdalenian sites
(after VALENTIN & PIGEOT in VALENTIN et al. 2000, 130).*

What kind of functions can be evidenced in Upper Palaeolithic settlements and with what type of fire can they be associated? Is it possible to evidence other functions than material ones? These are the questions we want to try to answer with the Verberie example.

Reconstructing the use of fire in a site requires going through several steps for identifying fire features and making hypotheses about their function. In a first step, it is necessary to evaluate the loss of data or their transformation by

Figure 8.2. Hearth D1 in the middle of the upper occupation floor (II.1).
The hearth is lined by stones which have replaced previous stones cracked by fire.
Around the hearth: an activity area with most flint tools; in front left: refuse area with
the beginning of a big dump; in the rear: an empty semicircular space which may mark
the location of a shelter (tent or hut). (Photograph F. Audouze).

Figure 8.3. Hearth M1 with its filling of cracked stones,
flint flakes and fragments of bones (level II.1); to the right,
a few cracked stones from the lining are still in situ (Photograph F. Audouze).

taphonomic processes. It then requires reconstructing a hearth's history from its building phase to its phases of use, reshaping or rebuilding, modification and abandonment. Microstratigraphic observations and conjoining cracked stones is an efficient way to learn about these transformations. In parallel, several types of analyses can be performed and have been performed on hearths of sites of the MBP:

- anthracology for informing about fuel;
- micromorphology, particularly useful for detecting bioturbation, and microparticles unseen at a macro-scale;
- geochemical analyses for getting information about maximum temperatures and types of fuel (vegetal or mixed vegetal and animal fuel); and
- experimentation for identifying characteristics of hearth morphology, fuel types and duration.

Not all analyses give positive results. The alluviation brought by repeated floods covering the valley floors resulted in a rare preservation of the MBP settlements but have also taken away some kind of evidences: At Verberie, the flooding took away charcoals and most traces of ochre. At Pincevent and Etiolles ochre remained, but charcoals are scarce and de-structured. However, it was possible to identify pine and birch/willow used as fuel, in agreement with environment analyses (sedimentology and malacology) which indicate an open landscape of cold steppe with few trees (THIEBAULT 1994, 118–119). Micromorphology confirms these results with a diversity of fuels: conifers, broad-leaved trees, herbaceous plants and bones in Pincevent (WATTEZ 1994, 120–127). At Verberie, micromorphology evidences an intense bioturbation, but also presence of microchips of burnt bones and microcharcoals in hearths. However, experimentation performed by R. MARCH proved that hearth morphology and type of fuel were independent from function before late prehistory, mostly depending on local resources in fuel. He also proved that, in many cases, temperatures obtained in hearths remained moderate (MARCH 1995 unpublished dissertation, 485–486; 1996).

As a whole, micromorphology analyses and thermic analyses indicate that Magdalenians were able to control both the heat and the fuel consumption which remained moderate in Verberie M20 hearth and some of Etiolles hearths, and of longer duration and more intense at Pincevent (WATTEZ 1994; MARCH 1995 unpublished dissertation). Thus, we have to look for other kind of evidences. Three other types of data that are not directly related to hearths are eventually essential in order to identify their function:

- analysis of the spatial context in which the hearth is located;
- microwear analysis of flint blades and tools which document the tasks performed in its vicinity; and
- ethnoarchaeological models which generate diagnosis characteristics for a particular function.

At Verberie, eight hearths and two ashes refuse areas were uncovered, with a maximum of three hearths in layer II.3 and two in layer II.1. Though large artefacts are clearly *in situ* with well defined spatial distribution, analysis of taphonomic processes by micromorphology evidences a heavy bioturbation due to worms and illuviation with the result that charcoals and ochre were washed away. It also proves that "esquilles" (flint and bone chips smaller than 1 cm long) have moved in an uncontrolled manner and cannot be used for microstratigraphy, particularly in hearths. Microchips of bones are present on all thin sections indicating that bone may have been used as a fuel in addition to vegetals. Last, loss of carbonates and loss of density in the sediment matrix (silt) seems to result in an otherwise unexplained discrepancy between Palaeolithic hearths and experimental hearths: ancient ones exhibit no rubefaction while modern ones settled in the local sediment do (MARCH; unpublished Verberie excavation report, 1993).

There exists a certain amount of variability among the Verberie eight hearths:

- Larger hearths have a round or slightly oval basin ranging from 50 to 70 cm in diameter and an average depth of 15 cm. They have a lining of big stones or a filling of small ones. Seven hearths belong to this category (*Figures 8.2 & 8.3*).
- A small hearth has a basin 25 cm wide and 10 cm deep (hearth K5).
- An unprecedently known type was seemingly uncovered in 2000: a small oblong and shallow pit 50 cm long, 10 cm wide and 10 cm deep, with a discontinuous filling of small stones (*Figure 8.4: hearth J10*). A year later, it turned into a regular hearth, 60 cm in diameter, 15 cm in depth with a filling of stones.

Conjoining of cracked stones proved that hearths with a lining of big stones and those with a filling of small stones were two stages of the same type of hearth. They evolve through time from a lining of big stones toward a filling of small stones. The rejuvenation which took place at hearth D1, lined with big stones for at least a second time, could be demonstrates because old cracked stones were tossed away in a near-by dump to the exception of a small piece remaining in the hearth and which could be conjoined with several of the big cracked fragments from the dump. Besides, the presence of hundreds of stones cracked by heat indicate that they played an important functional part in heating and cooking, most probably as a mean for boiling water or manufacturing grease by boiling articular ends of bones (BINFORD 1978, 158–159).

Figure 8.4. Hearth J10 in level II.5. It cannot be interpreted as the remains of a burnt log because of its depth, and its filling of little stones (Photograph F. Audouze).

Microstratigraphy, with location of all artefacts in 3 dimensions permits to identify several phases of use, particularly for hearth L7 and hearths 05/6 which were in use in two consecutive levels: II.3 and II.4. Charcoals are absent (as explained earlier) and micro-charcoals visible in thin sections are so fragmented that they cannot be identified. Screening with a 50 microns mesh showed tiny residues of burnt bones most probably coming from bones used as fuel. But we saw earlier that fuel and hearth morphology are not the essential keys to understanding Magdalenian hearths function.

If we turn to other types of data we first look to context: we find that most large hearths except one are located in the centre of activities areas. In these latter that draw a circle around the hearth, we find most of the flint tools, the backed bladelets and small fragments of bones. Large pieces of flint, stone or bone are found in dumps and around butchering activity areas. At Verberie, microwear analysis documents meat cutting, bone working, hide processing, tools and backed bladelets re-hafting (KEELEY 1987). Blades for cutting meat are located in different areas, but they are in greater density close to butchering activity areas. These have been identified according to BINFORD's Nunamiut model (BINFORD 1983, 120–124; 169–170) by empty spaces bordered with useless bones (reindeer vertebrae in connection, coccyx, sternum, carpsis and tarsis). On the reverse, burins, backed bladelets, piercers and micropiercers are prevailing in central activity areas and we can assume that central hearths are giving light and heat for domestic and craft uses such as cooking, knapping, and retooling. Whether there are shelters or not – which is still to be debated in Verberie –, activity areas are outside around the hearth.[2] Four of the five basin hearths belong to this category of domestic hearth. But the fifth one has a different context: no tools were in the immediate vicinity and it is adjacent to a bones and stones dump. We thus have to turn to another approach: to ethnoarchaeological models.

Ethnoarchaeology is not yet rich in models for hearths. However hearths for smoking hides are documented by the famous paper BINFORD published on "smudge pitts" (BINFORD, 1967) and by research performed by S. BEYRIES on the Black foot of Yukon and Inland Shumash of Northern Columbia (BEYRIES – PETREQUIN 2001). Combining the results of the two authors, we get the following characteristics: hearths for smoking hides are straight-sided shallow pits with essentially flat bottom in which carbonization of fuel is performed in a reduced atmosphere. The size of the pit is adapted to the size of the animal skin. The depth is important in order to avoid blaze and sparkles that could damage the skin.

One hearth at Verberie may be a candidate for such a function. In its first phase of use it is 50 cm by 40 cm wide with an unusual depth of 20 cm. (*Figure 8.5*). There is no activity area around it, no tools, as said earlier, only an empty space that may have been a butchering activity area. In its second phase however, it is wider and shallower and its centre is filled with a big stone, then, filled with refuse. Its function as smoking pit remains in question since several features contradict this interpretation: it has a lining of oblique or sub-vertical stones, use

---

[2] An oval or semi-circular empty space behind the activity area of hearth D1 in level II.1 can be considered as a shelter. It is bordered by activity areas and by a dump which has the characteristics of the door dump of Binford's model (BINFORD 1983, AUDOUZE 1987). At Etiolles, two occupation units have a shelter clearly defined with a circle of stones, with a central hearth inside. At Pincevent, shelters are located as at Verberie and the hearth is outside.

Figure 8.5. Hearth O5/O6 in level II.3 with its lining of big stones and a flat stone at the bottom. The hearth is not yet completely dug, a cube of dirt in the left rear being reserved for micromorphology analysis (Photograph F. Audouze).

of which does not clearly appear. Moreover, the screening of the filling yielded a much higher quantity of micro-chips of bone and flint than in other hearths (with 1 and 2 mm mesh). The presence of a big central stone was also observed in hearth L7 at the end of its earlier phase but we do not know if it is a kind of condemnation or a functional feature. The two stones seem thick and too dense to be used as cooking stones and were not removed when these hearths were rejuvenated for a second phase of use (*Figure 8.6: profile of hearth L7*).

The second well-documented category of non-domestic hearths is concerned with hearths for drying meats for preservation. Again, only one or two characteristics are essential: the hearth pits must be shallow and have a flat bottom. Their dimensions correspond to the size of the rack from which the pieces of meat are hanging. No such hearth can be found at Verberie but meat can be dried without smoking. A third category is well documented for sweat lodges with a filling of big stones but it does not seem to be relevant here, though the vast quantity of heated and cracked stones in dumps indicate that boiling water with heated stones was a common way of cooking.

The last two hearths have distinct features that do not let them to be classified in any known category. K4 is a small basin, 25 cm in diameter and 10 cm deep, filled with large bones and a flint core and prolonged by a shallow pit, 50 cm long, 30 cm wide and 8 cm deep. The latter is filled with fragments of large bones extremely well preserved but this may be a refuse corresponding to a secondary use and better preserved than elsewhere because of the high content of organic material in the sediment. When we turn on to the social and symbolic realm, no need to say that the task is much more difficult. Though such an essential component or prehistoric life such as fire could not be deprived of social and symbolic investments, it is very difficult to evidence them. Material culture only gives partial clues. Since we are not able to recognize the difference between men's tools from women's tools, we cannot interpret any different spatial distribution of tools and more generally any spatial organisation in relation to gender. We can only observe that in level II.1, tools around hearth D1 are predominantly burins and backed bladelets while tools around hearth M1 are predominantly piercers and micropiercers (AUDOUZE 1987, 1988).

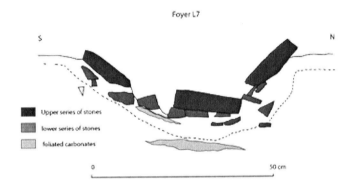

Figure 8.6. Profile of hearth L7 in level II.4 with its central stone.

We may find a spatial relation of flint knapping activities in relation to the age of knappers. At Verberie, as at other Magdalenian sites of the Paris basin (PIGEOT 1987; OLIVE 1988; PLOUX *et al.* 1991; OLIVE *et al.* 2000), the existence of knappers of different competence and knapping skills could be demonstrated. Unskilled knappers are obliged to prematurely abandon cores because of knapping mistakes that render the core useless. These damaged cores are usually uncovered further away from the domestic hearths than well knapped cores. We thus can assume than experienced knappers were the ones allowed to sit close to the hearth for their craft activities. Unskilled knappers were most probably youngsters since learning flint knapping is a long process requiring the acquisition of an extended know-how and motor ability. So, instead of a gender partition of space, what we find is a more flexible distribution according to age, very well evidenced at Etiolles. Once more, Palaeolithic women remain invisible.

Shifting to the symbolic realm or more modestly to the ideology realm is even more frustrating. Fire, essential to the survival of Palaeolithic hunters, must have been an essential element of the household symbolism. However, it may not play an important part beyond the household for people who chose to locate most of their sanctuaries in dark caves. There, fire is only present as torches lamps or hearths. And up to now, these torches, lamps and hearths seem to have played an essentially practical part for light and charcoal (CLOTTES 1995, 88).

The only possible element of symbolism I find in the Magdalenian of the Paris basin is the spread of ochre around the hearths. At Verberie it is only present as vestigial traces under flint tools or stones where it was preserved in carbonate crust. At Pincevent it is very well preserved and tints the sediment of activity areas around domestic hearths (LEROI-GOURHAN – BREZILLON 1972, 238–241). But we have no clue about its significance. We can only relate it to the ochre spread in some areas of caves (that Leroi-Gourhan assimilated to female zones of caves; LEROI-GOURHAN 1966/1992, 199). Ethnoarchaeology cannot provide us with significant models because cultural particularisms are much stronger in the symbolic realm than in material culture and prevent from finding any valid generalization. Thus, fire, absent from Palaeolithic art, does not permit us to break its secrets.

Bibliography

AUDOUZE F. 1987
   The Paris Basin in Magdalenian times. In: Soffer, O. (ed.), *The Old World Pleistocene*, New York, Plenum, 183–200.

AUDOUZE, F. 1988
   Des modèles et des faits: les modèles de A. Leroi-Gourhan et de L. Binford confrontés aux résultats récents. *Bulletin de la Société Préhistorique Française* 84, 343–352.

BEYRIES, S. 1997
   Systèmes techniques et stratégies alimentaires: l'exemple de deux groupes d'indiens de Colombie-Britannique. In: Patou-Mathis, M. (ed.), *L'alimentation des hommes du Paléolithique*. ERAUL No. 83, 73–92.

BEYRIES, S. – PETREQUIN, P. 2001
    Introduction. In: Beyries, S. – Petrequin, P. (eds.), *Ethno-archaeology and its transfer*, European Archaeologican Congress, Bournemouth (Great Britain), 14–19 September 1999. BAR International Series, 8–20.

BINFORD, L. 1967
    Smudge pits and hide smoking: the use of analogy in archaeological reasoning, *American Antiquity* 32 (1), 1–12.

BINFORD, L. 1978
    Nunamiut Ethnoarchaeology. New York, Academic Press.

BINFORD, L. 1983
    In pursuit of the past. Decoding the archaeological record. New York, Thames and Hudson.

CLOTTES, J. 1995
    Postface. In: Chauvet, J-M., Brunel Deschhamps, E. – Hillaire, C. (eds.), *La grotte Chauvet à Vallon Pont d'Arc. Paris*, Le Seuil (coll. Arts rupestres).

ENLOE, J. G. 1996
    Seasonality and age structure in remains of Rangifer tarandus: Magdalenian hunting strategy at Verberie. *Anthropozoologica* 23, Paris.

ENLOE, J. G. – DAVID, F. 1995
    Rangifer herd behaviour: seasonality of hunting in the Magdalenian of the Paris Basin. In: Jackson, L. J. – Thacker, P. (eds.), *Caribou and Reindeer Hunters of the Northern Hemisphere*, Avebury Press, 47–63.

KEELEY, L.H. 1987
    Hafting and retooling at Verberie. In: *La main et l'outil. Manches et emmanchements préhistoriques.* Table-ronde CNRS de Lyon 26–29 novembre 1984. Lyon, Ed. Maison de l'Orient/de Boccard (Travaux de la Maison de l'Orient n°15), 88–96.

LEROI-GOURHAN, A. 1966/1992
    Réflexions de méthode sur l'art paléolithique. *Bulletin de la Société Préhistorique Française*, LXIII. Republished in: Leroi-Gourhan, *A. L'art pariétal, langage de la Préhistoire.* Grenoble, Editions Jérômer Millon (Coll. L'homme des origines), 183–201.

LEROI-GOURHAN, A. – BRÉZILLON, M. 1972

Fouilles de Pincvevent. Essai d'analyse ethnographique d'un habitat magdalénien. Paris, Editions du CNRS (VIIe supplément à Gallia Préhistoire).

MARCH, R. J. 1995

Méthodes physiques et chimiques appliquées à l'étude des structures de combustion préhistoriques: l'approche par la chimie organique (Unpublished dissertation for the doctorate of the Université de Paris I Panthéon-Sorbonne, 1995).

MARCH, R. J. 1996

L'Etude des structures de combustion préhistoriques: une approche inter-disciplinaire. In: Bar Yosef, O., Cavalli-Sforza, L., March, R. J. – Piperno (eds.), *XIII International Congress of prehistoric and protohistoric sciences Forli-Italia*-8/14 September 1996. Colloquia 5: *The Lower and Middle Palaeolithic*, Colloquium IX, 251–75.

OLIVE, M. 1988

Une habitation magdalénienne d'Etiolles. L'unité P15. Paris (Mémoires de la Société Préhistorique Française – tome 20).

OLIVE, M. – AUDOUZE, F. – JULIEN, M. 2000

Nouvelles données concernant les campements magdaléniens du Bassin Parisien. In: Bodu, P., Kristiansen, M. – Valentin, B. (eds.), *L'Europe centrale et septentrionale au Tardiglaciaire.* Actes de la Table Ronde de Nemours 13–16 mai 1997, Nemours, Ed. APRAIF (Mémoires du Musée de Préhistoire d'Ile de France, 7), 289–304.

PIGEOT, N. 1987

Magdaléniens d'Etiolles. Economie de débitage et organisation sociale. Paris, Editions du CNRS.

PLOUX, S. – KARLIN, C. – BODU, P. 1991

D'une chaîne l'autre: normes et variations dans le débitage laminaire magdalénien, *Techniques & Culture*, 17–18, 81–113.

THIÉBAULT, S. 1994

Analyse anthracologique. In: Taborin, Y. (ed.), *Environnements et habitats magdaléniens dans le centre du Bassin parisien*, Paris, Editions de la MSH, (DAF n°43), 118–119.

VALENTIN, B. – PIGEOT, N. 2000

Eléments pour une chronologie des occupations magdaléniennes dans le Bassin parisien. In: Valentin, B., Bodu, P. – Christensen, M. (eds.), *L'Europe centrale et septentrionale au tardiglaciaire*. Actes de la Table-Ronde internationale de Nemours, 14–16 mai 1997. Nemours, Ed. APRAIF (Mémoires du Musée de Préhistoire d'Ile de France, 7), 129–138.

WATTEZ, J. 1994

Micromorphologie des foyers d'Etiolles, de Pincevent et de Verberie. In: Taborin, Y. (ed.), *Environnements et habitats magdaléniens dans le centre du Bassin parisien*, Paris, Editions de la MSH (DAF n°43), 120–127.

9

Keep the Yurt Fires Burning: Ethnographic Accounts and Religous Myths Surrounding Indigenous Fire Use in Western Siberia

ANN MARIE KROLL-LERNER

Introduction

The importance of fire in the daily lives of non-industrial people worldwide is incontrovertible. The domestic use of fire for cooking, lighting, heating, and defending forms a basic element in the survival of a household or community. For groups living in Siberia the life-saving hearth keeps the long, cruel winters at bay. The popular images of Siberia as a frozen "wasteland" are not without a basis in reality. Permafrost binds two-thirds of Siberia, in many areas up to one kilometer deep. Within the inhabited areas, temperatures drop to -70° Celsius freezing rivers and seaports for the majority of the year. In the face of these natural challenges, prehistoric and historic peoples survived and flourished for thousands of years. Fire has been used as a weapon against wild animals and human adversaries alike. There is ample archaeological and historical evidence for the raiding and burning of villages and crops.

But fire has more than pragmatic uses. For many western and central Siberian groups, fire is understandably a symbol of life, a terrestrial representation of the burning sun, and a purifier of the living and the dead. Through various rites of passage – birth, marriage, and death – fire is used for bringing the sacred and profane into discourse. For example, the hearth is a fundamental structure within a dwelling, the *yurt*, of Siberian nomadic and semi-nomadic people, providing light, heat, and cooking area while also providing the focus for ritual practices. This chapter examines these symbolic aspects of fire-use by indigenous nomads, special events or actions that are different from the daily uses of fire. By using ethnographic accounts in unison with related historical/religious texts, archaeologists can create "interpretive discovery" (CUNNINGHAM 2003, 391) – to identify sacred cultural practices of prehistoric populations by their archaeological signatures.

Pastoral Nomads of Siberia

Nomadic studies have long focused on groups living in the Middle East and Africa. Siberia remains another kind of *wasteland* – one where knowledge of pastoral peoples is expressed but not emphasized. In most social science texts, Siberia remains an empty space. This is neither a fault nor result of inattention – much has been written about this vast Eurasian landmass by Russian and Soviet scholars, political appointees, travelers, historians, and ethnographers. The accounts are largely inaccessible due to the obscurity of publication, language barriers, or political agents. Some historical texts (especially HERODOTUS' *Histories* [1995]) have been taken on a singular importance or central focus in the work of history-based archaeologists, so that even the names given to archaeological materials and prehistoric peoples have been taken from the pages of history (i.e. the Scythians, Masagatae, and Sarmatians). Religious texts, in particular the *Rig Veda* and *Avesta*, are also used to link archaeological traces to Indo-European or Indo-Iranian peoples, while conversely the archaeological remains have been used to reconstruct the religious practices of Indo-European groups (FOLTZ 1999). All of these historical texts are legitimate sources for comparison and interpretation. These lines of evidence can and should also be coupled with living populations (or recently extinct ones) represented in the work of ethnographers.

Archaeologists recognize to the value of ethnography for analogous interpretation (GOULD – WATSON 1982; BINFORD 1983; WYLIE 1985) – not to create indelible laws but rather as explanatory hypotheses to describe the possible relationship between human behaviour and the material cultural remains. Archaeologists created an entirely new branch of archaeology, *ethnoarchaeology*, based on ethnographic data and experimental archaeology to model past activities. Archaeologists who were specifically interested in the remains of Eurasia are finding new insights into the lives of pastoral nomads – seasonal exploitation and migration (CRIBB 1991), human labor practices (KORYAKOVA – SERGEYEV 1995), the overlap of agriculture and pastoralism (CHANG – TOURTELLOTTE 1998). As the volume edited by BAR-YOSEF and KHAZANOV (1992) points out, we can no longer make sterile statements about the archaeological record when so much more can be said by using integrative approaches. Economic forces (pastoralism) should not be viewed in absence of social and ideological factors (CHANG 1993). By using living populations as our models in addition to historical texts, we construct more nuance interpretations. While heeding the dangers of using a direct historical approach to recreate past lifeways, ethnographic materials provide invaluable interpretive avenues into the archaeological record, especially in the realm of prehistoric behaviour patterns.

Historic Texts and Ethnographic Groups

The historic and contemporary ethnographic groups of western Siberia are made up of a melding of many different cultural groups moving through and into the grasslands. Archaeologists commonly accept that during the Bronze Age inhabitants of the Eurasian steppe spoke Indo-Iranian and eastern Iranian languages (MALLORY 1992) while mixing with the Finno-Ugric speakers living in the taiga (KUZMINA 1994). Thus by the Iron Age the ecological zone in between, the forest-steppe, represented an interaction sphere of these two linguistic groups (MATVEEYEVA 2000), creating a *culturetone* of greater linguistic and cultural diversity than seen in either the steppe or taiga individually. Much as an ecotone, the area at the intersection between two different ecological zones, reveals greater biological diversity, the same can be said for the area between two cultural groups, what I term a culturetone.

With the 'Great Folk movement' of the first millennium AD, Turkic language groups began replacing the Iranian ones (KORYAKOVA 1998, 209). Thus when considering the use of ethnographic data, often the "lines" dividing protohistoric and historic populations become blurred (at best). For example, the Yukagir of northeastern Siberia are Palaeo-Asiatic speakers living in the tundra and steppe, whose language has close ties to Uralic-Altaic. And while the Khanty/Mansi of the Ob River basin are Finno-Ugrian nomadic reindeer herders, historians attribute their origins to the "Yugrians" discussed in the 12th century *Chronicles*, or originating from a mixture of peoples east of the Urals – a steppe-Uralic blending (OLSON 1994, 377). A discussion of diffusion, assimilation, and acculturation might be appropriate, but is not the focus of the current chapter (but is most thoroughly done in RENFREW's 1990 book, *Archaeology & Language: The Puzzle of Indo-European Origins*).

At the time of Russian conquest there were as many as 120 indigenous languages or dialects in Siberia – a number that may represent a vast oversimplification of the difference between incomprehensible languages and local variants of the same language, or *dialect* (FORSYTH 1991, 70). Iranian and Turkic speakers, coming from two distinctive language families (Indo-European and Uralic-Altaic respectively) nonetheless had many cultural and historical similarities that make them less distinguishable in the archaeological record. While inhumation in barrows (*kurgans*), and the use of wheeled vehicles is evident, horse nomadism is the primary link between them. Archaeologists recognize that the horse was likely domesticated somewhere on the Eurasian steppe by Indo-European speakers (the debate that rages between riding or eating horses, and/or using horses for traction can be read in ANTHONY 1986, 1995; ANTHONY – BROWN 1991, 2000; LEVINE 1998, 1999, LEVINE *et al.* 1999; and OLSEN 2000). The religious practices of Indo-Europeans, put forth by George Dumézil and James Mallory, are also observable among Uralic-Altaic peoples (FOLTZ 1999, 25) including the universal dominion granted by a sky god, and fire worship. It is because of these historic and cultural commonalties that the religious texts of the Indo-Europeans, as well as ethnographic examples from Uralic-Altaic groups are being combined. The nature of their mobility and economic adaptation, nomadic pastoralism, created the situation where sharing, mixing, conquering, and adopting would be possible if not probable.

Prolific accounts of travelers, government officials, map-makers, and "ethnographers" (not always self-defined as such, but nonetheless using their personal insights to describe the indigenous groups) exist for certain areas within Siberia at certain times. The early Kievan Chronicles of the eleventh and twelfth centuries mention "Ugra" or "Yugrians" living in the eastern Ural foothills. Plano Carpini's 13th century missions to the Mongols was a widely read and popular text; map-mapkers were sent to the Transurals region (Tobol River) and reported on the state of affairs in the Tsar's realm (Remezov and his son during the late 17th and early 18th century). By the 18th century descriptions by foreign travelers – Gmelin (German in Siberia 1733–1743), STRAHLENBERG (Swede describing "Tartars", antiquities, monuments, and geography in 1738), Müller (officer of the tsar in 1787 "Ostiak" and "Vogul" or Khanty and Mansi along the Irtysh and Ob rivers – not to be confused with Max Müller who wrote *People of Siberia* in the mid-19th century) – and the Russians – Georgi (southern Siberia 1773–1774), Levshin (published on Kazkh Turks in 1832), and Verbitskii (describes Turkish shamans in Altai in 1840) – all these recount various curiosities and provoked early interest in Siberian peoples.

Siberia of the 17th and 18th centuries was a place of indigenous conflict, Russian military governance, and aggression. It housed a refuge for political dissidents and religious refugees, and criminals (BLACK 1991, 60; COLLINS 1991, 37). Gross description was often not the focus of these early works and one is lucky to glean small components of the individual cultures. Peter the Great did order the exploration and documentation of natural and human resources of western and central Siberia, but much of this effort resulted in the looting of archaeological sites, the proceeds of which returned to the Hermitage in St. Petersburg. During the course of these and other administrative trips, indigenous people were counted and described in order to add them to the tax rolls. It was noted during the late 16th–early 17th century that the native population was well-suited to mounted battle, and a real "thorn in the Russians' flesh for over a century":

> *In the cavalry skirmishes against Kalmyk, Krigiz, or Mongol horsemen, who almost lived in the saddle and were a 'numerous, highly maneuverable and well armed foe, a serious and dangerous enemy', Russian firepower proved of very little advantage.*

(Nikitin 1987, 53–6 cited in COLLINS 1991, 44)

By the 19th century the landscape of ethnography had changed. Scholars, recognizing the imminent loss of Siberian cultures to Russian "immigration" (sometimes voluntary, but as likely a result of the relocation of state criminals), endeavored to record the cultures of these groups before they were lost altogether. It is therefore during this period that the most clearly anthropological of the cultural descriptions is to be found. Within the ethnographic work of the 19th and 20th centuries I further restrict my examples to only those of the Indo-European and Uralic-Altaic speaking pastoral nomads, those with roots in my region of interest, and thus those "more likely analogous to past activities" (Watson cited in GOULD – WATSON 1982, 359) of the archaeological cultures from the late Bronze and early Iron ages. Within this limited sample the ethnographic sample is further constrained by discussions of fire.

Ethnographic Record tuned by Religious Texts

Two primary ethnographic sources provide the vast majority of references to indigenous fire use by western Siberian pastoral nomads – the first, a collection of ethnographic works originally published during the 19th and 20th centuries but edited into an ethnology written by M. A. CZAPLICKA (1914); and the second, a historical-ethnographic survey edited and written during the Soviet era by LEVIN and POTAPOV (1956). Each of these tomes broadly surveys both in a geographical sense as well as in the number of cultural groups included. Specific ethnographies were used to supplement the discussion of individual groups where needed. The list of groups (*Table 9.1*) is only a portion of those included in the conference paper upon which this chapter is based. Upon the advice of John Chapman, the discussant in our session, I have focused my current discussion on only those groups who meet specific criteria. Linguistically, for reasons stated previously, they must come from either the Indo-European or Uralic-Altaic families. Each group must be nomadic or semi-nomadic, and have a pastoral economy that focuses specifically on horse, cattle, and sheep/goat. Geographically they must originate in or historically reside within west/central Siberia – as these are nomadic groups, their range may include other areas. Occasionally groups who are/were nomadic have been included if multiple aspects of their cultures overlap with the above criteria. These restrictions are related to the archaeological cultures around which my own research revolves. The cultural groups I focus on are as follows: Buryat, Bashkir, Krygyz, Tatar, Yakut, and Khanty/Ostyak.

Table 9.1 Ethnographic Groups (From J. S. OLSON 1994; J. FORSYTH 1992).

Group	Language Family	Language Branch	Geographic Location	Ecological Zone	Economy*
Ossetian/ Alans	Indo-European	Iranian	Caucasus	Mountains, foothills, valleys	Farming
Buryat	Altaic	Mongolian	Irkutsk, Transbaikal	Steppe	*PN*–horse, sheep/ goat, cattle
Altai Bashkir Kyrgyz Tatar Yakut	Altaic	Turkic	W. Siberia, Central Asia Altai	Steppe, Forest-steppe	*PN*–horse, sheep/ goat, cattle
Khanty/ Ostyak	Uralic	Ugric	Urals, Ob River to Far East**	Arctic tundra/ taiga	*Nomadic–reindeer, hunting/ fishing*
Enets/Nenets (Ent)	Uralic	Samoyed	Lower Ob and Yenesei	Arctic tundra/ taiga	*Nomadic–reindeer, hunting/ fishing*
Yukagir	Palaeo-Asiatic (Uralic-Altaic) ***	Yukagir	Northeastern Siberia	Tundra and steppe	*Hunting/fishing; few reindeer herders*

*   Traditional location & economy prior to Soviet collectivization
**   Historical debate places Khanty origins in northeastern European Russia, the Urals, or Irtysh River basin
*** Yukagir has close ties to Uralic and Altaic languages probably due to considerable mixing

PN = pastoral nomadism

Fire-worship is a long-standing practice there. Neither the old folks nor the
young would forget to feed the fire, to throw pieces of food into it and thus win
its favor. Incidentally, they don't forget nowadays either. Young people do it
seemingly as a joke, but they still do it, keeping the superstition alive for the
time when science will fail...

VLADIMIR RASPUTIN's *Siberia, Siberia* (1996, 326)

This passage by the poet Vladimir Rasputin shows the 20th century continuation
of fire's ritual symbolism, even if the younger generation maintains the ritual for
"superstition's" sake. Either through complete adherence to a set of rules or as an
indicator of the relentless motion of culture, rituals continue through generations
often times with only slight modifications to meaning or stages. To draw analogies
from fire worship among ethnohistoric cultures to material remains of fire in
the archaeological record examples are drawn to help establish the relationship
between these lines of evidence separated by time and space.

Generations passed the *Avesta* and the *Rig Veda* through oral tradition, some
suggest for nearly a millennium (while Zoroaster (Greek)/Zarathushtra, the
prophet and author of the earliest hymns, the *Gathas*, and around whose teachings
the religious practices related in the *Avesta* are based, lived somewhere between
the 13th and 6th century BC, the *Avesta* was not written down until the 3rd century
AD; the Vedic hymns have portions dating to 13th century BC, but was composed
during the 6th century BC – dates approximately agreed upon, except for the dating
of the *Rig Veda*, which RENFREW states was "first set in writing as late as the 14th
century AD" (1990, 12)). Both texts consist of Iranian dialects, with the *Avesta*
so strongly resembling Vedic Sanskrit as to reveal a "protolinguistic ancestor"
(PUHVEL 1987, 38), Indo-Iranian. Each teaches the lessons of the gods. While
Zoroaster broke with his own warrior class and opposed tendencies common
amongst his Indo-European peers (various sacrifices and drinking rituals), he
professed a dualistic form of monotheism – *Ahura Mazda (or Auramazda)*, the
god of the sun, and *Angra Mainyu*, the evil divinity. Ahura Mazda is reached
through the medium of fire, good thoughts and deeds:

We therefore bow before Thee, and we direct our prayers to Thee with
confessions of our guilt, O Ahura Mazda! with all the good thoughts (which
Thou dost inspire), with all the words well said, and the deeds well done, with
these would we approach Thee. And to Thy most beauteous body do we make
our deep acknowledgments, O Ahura Mazda! to those stars (which are Thy
body); and to that one, the highest of the high, [such as the sun was called]!
Book of Yasna 36. 5–6

The sun and its earthly manifestation, fire, are the visible forms of Ahura Mazda for many people after Zoroaster, including the Assyrian's, *Asara Mazas*, and the Saka, whose word for sun was *urmaysde* (HARMATTA 1994, 315).

The *Rig Veda* directly addressed or spoke of gods and goddesses, with more than 200 of its thousand hymns devoted to *Agni*, the god of fire. He is represented by flame and smoke in sacrificial fires, the hearth, cremation fires, lightning, and the sun itself (O'FLAHERTY 1981). As the sun or the sun-bird, Agni represents the birth of the new year and daily rejuvenation (*Rig Veda* 10.123, 10.177), he is born of the sky and earth, but also of the waters (*R. V.* 1.164.52), and represented by the cow, he is the earth itself – thus reinforcing the Indo-European concept of cyclical time, he is the mother and the son. Agni is both the herdsman (*R. V.* 1.164.21) and the horse, fashioned out of the sun (*R. V.* 1.163.2), conveying the dead (*R. V.* 1.26.1, 5.2.1, 10.51.6), his mane as rays of the sun (*R. V.* 1.163.11) that parallel his fiery flames resembling long hair (*R. V.* 1.164.44). Agni is seen in the oblation fire as smoke, in the domestic fire, he is fed fat ("brother with butter on his back" *R. V.* 1.16.1).

Historic nomadic groups worshipped the sun and fire. In 521 BC, on the Bahistan Inscription, Darius, the Achaemenid (Persian) leader, wrote that he received his throne from *Auramazda*, smote many enemies and subjugated their peoples (TOLMAN 1908), but also denounced the nomadic Saka because they did not worship the sun god (FOLTZ 1999). This may have been his way of belittling Saka beliefs and practices, a simple slur. The Indo-Iranian cultural affinities, i.e. horse nomadism, barrow-building, etc. – of the Saka and linguistic reconstructions place them as the easternmost of the Iranian speakers, and Herodotus relates the story of the Massagatae queen who swears an oath "by the Sun our Master" to whom the Saka sacrifice horses – the swiftest animal to the swiftest god (*Histories* II, 215–216). The Huns too sacrificed horses and other animals to the gods of fire, "all creatures in their eyes considered remarkable" (BALDICK 2000, 30); Tungus, Buryat, Yakut, and other Turkic shamans communicated with the animal spirits using fire as the messenger (FORSYTH 1991; LEVIN – POTAPOV 1956). While the Khanty (Ostiak) used fire to dedicate sacrificed reindeer to the ancestral spirits (BALZER 1999, 11), the Mongols sacrificed butter to the hearth fire in request that in the new year the fire-deity, here Queen Mother, might bring more sons, brides, and sons-in-law (BALDICK 2000, 117). Changing cultural influences, including new religious and political ideologies, came into the lives of indigenous Siberians, but as CZAPLICKA noted in 1914, and as is seen into the 21st century, the cult of fire remains.

Higher or formal prayer and worship was to be performed by a Zoroastrian priest in the fire temple, but the daily spiritual needs could be fulfilled before the household fire where both *Ahura Mazda* and the Vedic *Agni* were located. The power of this domestic fire worship was noted by Engels, who witnessed Siberian peasants purifying areas using smoke and flames, while folk doctors cured everything from rashes to herpes by conjuring, "Fire, fire, take your fire from God's slave (say name). I have been beyond the sea and put out the fire". (BOLONEV 1992, 79) The Buryats made sacrifices of oil, fat, and meat to the hearth (VYATKINA 1956, 227) – similar to the feeding of the Vedic domestic fire (*R.V.* 1.164.21). Scythians reportedly bathed not in water, but cleansed their bodies in smoke and ash (Herodotus *Histories*), a purification custom carried on by Mongolian and Turkish peoples. These common themes of fire as messenger, purifier, and god to be worshiped are seen even more clearly in the rites of passage of pastoral nomads.

Rites of Passage

The *Avesta*, in particular the earliest Gathic sections, is largely silent on birth and death rituals; Zarathushtra was clear that ritual was not religion, another break with the traditions of his day. The *Rig Veda* by contrast is replete with ritual references. West and central Siberian pastoral nomads' use of fire most significantly resembles the rites described in the *Rig Veda* during ceremonies commemorating significant life changes. Rites of passage (VAN GENNEP 1960) are transformative, removing an individual from one position (*separating*), creating a vehicle of change (*transitioning*), and reconnecting the individual with his/her society in a new or altered way (*incorporating*). Each occasion may emphasize one or another of the three stages of transformation to a varying degree, and as will be shown in the three primary occasions marked by fire ceremonies –birth, marriage, and death – each has a varying degree of relevance to an archaeologist. Each of these ceremonies is viewed in the ethnographic context and then compared to the Indo-European religious texts.

Birth and Marriage

The cultural similarities between the nomads of west/central Siberia are significantly expressed during rituals surrounding the birth and naming of individuals and marriages. Physical birth and the naming rituals are often nearly simultaneous (at least within several days), as there generally is no recognition of personhood until a name is given, while the continuity of the birth rite of passage might be considered to stretch from conception – separating the seed/sperm (seen as an element of fire in the *Rig Veda*) from the male, "the seed fire is placed in the water womb" (O'FLAHERTY 1981, 26). Or perhaps separation begins with the child leaving the womb. The vehicle for changing the baby into a member of Buryat society involves a feast (slaughter of a ram or cow) and a wrapping ceremony where questions are asked of the community/family – should the child be wrapped or should the sacrificed cow bone (usual reply is "the child") – each question asked three times. The feast ends with a cleansing fire being built at the location of the birth; oil – a substance squirted from the mouths of participants of the ceremony into the fire (CZAPLICKA 1914, 138–139). The swaddled child, now named, has been incorporated as a new member of the Buryat society. For the Yakut the birth ceremony is very similar in that they slaughter animals for food and for sacrifice to the fire. Here it is specifically a female divinity, *Ayisit*, who receives fat on the fire to ensure future gifts of children (*ibid.* 143).

As we see in several nomadic groups, marriage begins with wife "stealing" (Buryat, Khanty, Mongol and Altaian), where the bride is separated from her kin by use of real or exaggerated force. Once her doom is decided, fire becomes an important component in determining the outcome of the ceremony. A Yakut woman approaches the yurt of her husband's family bringing wood from the hearth of her former home. As she adds fuel to her in-law's fire, she pronounces herself "mistress, come to rule the hearth (CZAPLICKA 1914, 110–111), after which she quickly bows to her father-in-law (VYATKINA 1956, 225); within the confines of her new household, she may never pass between the fire and any male clan elder. Bowing before the flames can be seen as the respect given to Agni in the *Rig Veda*. The marriage house itself will be blessed if a Teleut (southern Altai) man can create a new fire without the use of coals from a former abode. After the two have lived in the house (*odakh*) for three days, it is torn down, but no fire can be made from the birch timbers. For the Buryat the conclusion of the marriage ceremony simply occurs when the groom, thrice summoned from the yurt, throws fat on a fire (*ibid.* 119). The greatest problem then becomes the hope for a fruitful union, as the greatest fear of a new couple is that there will be no children, and the "fire of the hearth will go out". This forms the basis for their strongest sworn oath, "May my fire be extinguished" (*ibid.* 134), in essence a lie would be tantamount to sterility, which directly parallels sentiments expressed in the *Avesta*, where extinguishing a fire is punished and repented (*Book of Yasna* 36).

The ethnographer or historian records births and marriages, and through the written descriptions of the uses of fire one recognizes its importance in these rites. But it is virtually impossible for an archaeologist to distinguish a fire with fat residues miraculously intact and recoverable from a common cooking hearth. These ritual activities taking place within households may prove beyond our grasp. It is really only in death ceremonies that the archaeologist is completely capable of recognizing the specific ritual.

Death

Morris states, "The biological death of an individual sets off a more prolonged social process of dying" (MORRIS 1992, 10), including the removal of an individual from amongst the living to become one of the ancestors. The existence of a body, a grave, or even a barrow does not represent the complete rite – we cannot use these archaeological traces to reconstruct the complexities of a culture's funerary ritual. Fire and the historical texts can begin to approach death in a more comprehensive manner.

I return to the *Rig Veda* for its rather thorough discussion of the role of Agni in the funeral pyre. It is Agni's role to mediate between the living and the gods – to mediate (*R.V.* 1.26) and to bring the gods to the sacrifice (*R.V.* 1.1). As god of the cremation fire, it is asked, "Do not burn him entirely, Agni, or engulf him in your flames. Do not consume his skin or his flesh. When you have cooked him perfectly, O knower of creatures, only then send him forth to the fathers" (*R.V.* 10.16.1), thus the corpse is purified, before being transported to heaven. If Agni consumes the body (overcooked), then it is a basal act, a preparation for animals; if he cooks the corpse gently, then he has cleansed the dead for the fathers. Within the cremation fire there are sacrificial animals – a goat (*R.V.* 10.16.4) as Agni's share, and fat/suet and cow limbs (*R.V.* 10.16.7) as protection for the corpse against the sometimes cannibalistic tendencies of the fire god (O'FLAHERTY 1981, 47). Finally the "transportation" comes in the form of a horse (*R.V.*10.56.6), "let your body, carrying a body, bring blessings to us and safety to you" (*R.V.*10.56.2). In this final passage, Agni propels the corpse to heaven, sanctifies the living, and renews himself.

The key elements of the Vedic hymns on death are the impurities of the dead, the animal sacrifices necessary to protect the dead and honor the gods, and fire's role in separating the living from the fathers. These elements have been shown in the birth and marriage ceremonies, but come together ultimately in the death rituals. Fire is for the Kyrgyz the "purest of things and makes everything else pure: notably it purifies the dead of filth and sin" (BALDICK 2000, 49). Fire is used to purify the body (Yakut, Altai and Buryat), horses are sacrificed to feed the gravediggers (Yakut), or placed with the deceased (Buryat and Altai) for their consumption or for the god's. The burial of a Buryat shaman, who is symbolized in life by copper disks representing the sun (VYATKINA 1956, 227), is described in CZAPLICKA (1914). For the Buryat, the corpse of a shaman is either cremated or exposed on a platform. While cremation reflects the Vedic tradition, exposure to the sun (celestial fire) is common among Iranian peoples. During three days of ceremonies, the shaman's body is purified by the burning of herbs, and taken to the cemetery on horseback. When burnt, the horse is sacrificed, a pyre is built, and the mourners leave the place. On the third day, what bones remain are collected, placed into a birch box and buried, or tucked into a tree. In the three days between death and cremation, it is believed that the shaman roams about his house, in a liminal state between the living and the dead. It is the fire that opens the door between living and dead (similarly revealed in Khanty folklore, BALZER 1999, 179), and ultimately transports his soul to heaven, where the ancestors have been waiting for him. Thus the shaman was purified by smoke, accompanied by a horse, burned by fire, and finally able to reach the fathers – the path that the Rig Veda prescribes.

Meanings in Archaeology

Much of the archaeological literature of the Eurasian steppe and forest-steppe nomads is rife with references to the "ritual use of fire" and rightfully so as there is much evidence recovered from mortuary contexts. At times historical similarities of the "fire-cult" are drawn from proto-historic populations to their Bronze Age or Neolithic forebearers (GENING 1977, 70; GRACH 1980; BASILOV 1989; DAVIS-KIMBALL *et al.* 1995; KORYAKOVA 1996; IATSENKO 1999), or simply put the important role of fire and animal sacrifice in mortuary rituals (KORYAKOVA – DAIRE 2000, 68). As HANKS (2003, 106, 111, 283) points out, these are the conventional approaches to the interpretation of mortuary sites. The *Rig Veda*, *Avesta*, and historical accounts (especially Herodotus), have been queried in light of the archaeological remains (KUZMINA 1982, 1994; YABLONSKY 1995; MATVEYEVA 2000), and here usually to recreate "ethnic" or "genetic" ties to historic populations. Myths, deities, symbols and rituals should be understood in local contexts, as created by elements drawn from a common, cultural ancestral pool.

My purpose has been to show the historical similarities based on economy, mobility, burial practice, and language in order that future archaeological interpretations might be greatly enlivened by the use of ethnographic materials and relevant historical texts. None of these elements can be used alone in the recreation of past societies. There are great dangers in lumping groups under one "ethnic" classification or cultural name, since many groups participated in different economic practices, spoke dissimilar dialects, had differing family (clan) loyalties, and ultimately moved and mixed with each other across the Eurasian landmass. I endeavored to use only the specific ethnographic and historical information from fairly similar contexts for this reason.

Linguistic similarities cannot be used as the sole means by which populations are seen to acquire new cultural motifs. Language provides a fair marker for diffusion at one end of the spectrum and full-blown migration at the other end, but it remains only one aspect of cultural identity. Kinship systems, religious practice, political organization, even physical characteristics can and should be employed to fully understand the relatedness of populations. In this venue fire is one avenue for exploring the cultural continuity of Indo-Iranian myths, and has created an "interesting" avenue for interpretation of archaeological remains. It is by no means BINFORD's "Rosetta Stone" (1983, 24), but perhaps a glyph.

Bibliography

ANTHONY, D. W. 1986
    The 'Kurgan Culture,' Indo-European Origins, and the Domestication of the Horse: A Reconsideration. *Current Anthropology* 27(4), 291–313.

ANTHONY, D. W. 1995
    Horse, wagon and chariot: Indo-European languages and archaeology. *Antiquity* 69, 554–565.

ANTHONY, D. W. – BROWN, D. R. 1991
    The origins of horseback riding. *Antiquity* 65(246), 22–38.

ANTHONY, D. W. – BROWN, D. R. 2000
    Eneolithic horse exploitation in the Eurasian steppes: diet, ritual and riding. *Antiquity* 74, 75–86.

BALDICK, J. 2000
Animal and Shaman: Ancient Religions of Central Asia. London, I. B. Tauris & Co, Ltd.

BALZER, M. M. 1999
The Tenacity of Ethnicity: A Siberian Saga in Global Perspective. Princeton, Princeton University Press.

BAR-YOSEF, O. – KHAZANOV, A. 1992
Pastoralism in the Levant: Archaeological Materials in Anthropological Perspectives, Monographs in World Archaeology, Vol. 10. Madison, WI, Prehistory Press.

BASILOV, V. N. 1989 (ed.)
Nomads of Eurasia. Los Angeles, Natural History Museum of Los Angeles County.

BINFORD, L. R. 1983
In pursuit of the past: decoding the archaeological record. New York, Academic Press.

BLACK, J. L. 1991
Opening up Siberia: Russia's 'window on the East'. In: A. Wood (ed.), *The History of Siberia: From Russian Conquest to Revolution*, 57–68. London: Routledge.

BOLONEV, F. F. 1992
Archaic elements in the charms of the Russian population of Siberia. In: Balzer, M. M. (ed.), *Russian Traditional Culture: Religion, Gender, and Customary Law*, London, M. E. Sharpe, 71–84.

CHANG, C. 1993
Pastoral Transhumance in the Southern Balkans as a Social Ideology: Ethnoarchaeological Research in Northern Greece. *American Anthropologist* 95(3), 687–703.

CHANG C. – TOURTELLOTTE, P. A. 1998
The Role of Agro-Pastoralism in the Evolution of Steppe Culture. In: Mair, V. H. (ed.), *The Bronze and Early Iron Age Peoples of Eastern Central Asia, Volume I: Archaeology, Migration and Nomadism, Linguistics*, Washington DC, Institute for the Study of Man, 264–279.

COLLINS, D. N. 1991
Subjugation and settlement in seventeenth and eighteenth-century Siberia. In : Wood, A. (ed.), *The History of Siberia: From Russian Conquest to Revolution*, London, Routledge, 37–56.

CRIBB, R. 1991
Nomads in Archaeology. Cambridge: Cambridge University Press.

CUNNINGHAM, J. J. 2003
Transcending the "Obnoxious Spectator": a case for processsual pluralism in ethnoarchaeology. *Journal of Anthropological Archaeology* 22(4), 389–410.

CZAPLICKA, M. A. 1914
Aboriginal Siberia, a study in social anthropology. Oxford, Clarendon Press.

DAVIS-KIMBALL, J. – BASHILOV, V. A. – YABLONSKY, L. T. 1995 (eds.)
Nomads of the Eurasian Steppes in the Early Iron Age. Berkeley, CA, Zinat Press.

FOLTZ, R. C. 1999
Religions of the Silk Road: Overland Trade and Cultural Exchange from Antiquity to the Fifteenth Century. New York, St. Martin's Press.

FORSYTH, J. 1991
The Siberian Native peoples before and after Russian conquest. In: Wood, A. (ed.), *The History of Siberia: From Russian Conquest to Revolution*, London, Routledge, 69–91.

FORSYTH, J. 1992
A History of the Peoples of Siberia. Cambridge, Cambridge University Press.

GENING, V. F. 1977
Sintashta Cemetery and the Problem of the Early Indo-Iranian Tribes (Molgil'nik Sintashta I problema rannik indoiranskikh). SA 4.

GOULD, R. A. – WATSON, P. J. 1982
A dialogue on the meaning and use of analogy in ethnoarchaeological reasoning. *Journal of Anthropological Archaeology* 1(4), 355–81.

GRACH, A. D. 1980
Ancient Nomads in Central Asia (Drevnie kochevniki v tsentre Azii). Moscow, Nauka.

HANKS, B. K. 2003
Human-Animal Relationships in the Eurasian Steppe Iron Age: An Exploration into Social and Ideological Change. Unpublished Ph.D. dissertation: Cambridge University.

HARMATTA, J. 1994
Religions in the Kushan Empire. In: *History of Civilizations of Central Asia,* Vol. 2. Paris, UNESCO.

HERODOTUS, 1995
Histories. Translated by Godley, A. D. Cambridge, MA, Harvard University Press.

IATSENKO, V. 1999
Sarmatian Funeral and Memorial Rites and Ossetian Ethnogarphy. *Anthropology and Archaeology of Eurasia* 38(1), 60–72.

KORYAKOVA, L. 1996
Social Trends in temperate Eurasia during the second and first millennia BC. *Journal of European Archaeology* 4, 243–280.

KORYAKOVA, L. 1998
Cultural relationships in North – Central Eurasia. In: Blench, R. – Spriggs, M. (eds.), *Archaeology and Language II: Correlating archaeological and linguistic hypotheses,* 209–219.

KORYAKOVA, L. N. – DAIRE, M-Y. 2000
Burials and Settlements at the Eurasian Crossroads: Joint Franco-Russian Project. In: Davis-Kimball, J., Murphy, E. M., Koryakova, L. –Yablonsky L. T. (eds.), *Kurgans, Ritual Sites, and Settlements Eurasian Bronze and Iron Age.* BAR International Series 890, Oxford, Archaeopress, 63–69.

KORYAKOVA, L. N. – SERGEYEV, A. S. 1995
The Use of Ethnographic Models in the Reconstruction of Social and Economic Patterns (with reference to materials from research undertaken in Siberia). *World Archaeological Bulletin* (8), 18–29.

KUZMINA, E. E. 1982
On the Semantics of the Images on the Chertomulk Vase. *Soviet Anthropology and Archaeology* 21(1–2), 120–138.

KUZMINA, E. E. 1994
Where had the Indo-Aryans come from? The Material Culture of the Andronovo Tribes and the Origins of the Indo-Aryans (Otkuda prishli indoarii?). Moscow, Nauka.

LEVIN, M. G. – POTAPOV, L. P. 1956 (eds.)
The Peoples of Siberia. Chicago, University of Chicago Press.

LEVINE, M. A 1998
Eating horses: the evolutionary significance of hippophagy. *Antiquity* 72, 90–100.

LEVINE, M.A. 1999
Botai and the Origins of Horse Domestication. *Journal of Anthropological Archaeology* 18(1), 29–78.

LEVINE, M. – RASSAMAKIN, Y. – KISLENKO, A. – TATARINTSEVA, K. 1999
Late prehistoric exploitation of the Eurasian steppe. McDonald Institute Monographs. Cambridge: Oxbow Books.

MALLORY, J. P. 1992
In Search of the Indo-Europeans: Language, Archaeology and Myth. London, Thames and Hudson.

MATVEYEVA, N. P. 2000
Social and Economic Structures of the Western Siberian Population in the Early Iron Age (Sotsial'no-ekonomicheskiye strukturi naceleniya zapadnoi cibiri v rannem zheleznom veke). Novosibirsk, Nauka.

MORRIS, I. 1992
Death Ritual and Social Structure in Classical Antiquity. Cambridge, Cambridge University Press.

O'FLAHERTY, W. D. 1981 (translator)
The Rig Veda. London, Penguin Books.

OLSEN, S. L. 2000
Reflections of Ritual Behavior at Botai, Kazakhstan. In: Jones-Bley, K., Huld, M. E. – Volpe, A. D. (eds.), *Proceedings of the Eleventh Annual UCLA Indo-European Conference*, JIES Monograph No. 35, Washington DC, Institute for the Study of Man, 182–207.

OLSON, J. S. 1994
An Ethnohistorical Dictionary of the Russian and Soviet Empires. Westport, CT, Greenwood Press.

PUHVEL, J. 1987
Comparative Mythology. Baltimore, Johns Hopkins University Press.

RASPUTIN, V. 1996
Siberia, Siberia. Evanston, IL, Northwestern University Press.

RENFREW, C. 1990
Archaeology & Language: The Puzzle of Indo-European Origins. Cambridge, Cambridge University Press.

STRAHLENBERG, P. J. VON. 1970
Russia, Siberia, and Great Tartary. Reprinted from the original English translation (London 1738). New York, Arno Press.

TOLMAN, H. C. 1908 (translator)
The Behistan Inscription of King Darius. Nashville, Tennessee, Vanderbilt University Press.

VAN GENNEP, A. 1960
The Rites of Passage. Chicago, University of Chicago Press.

VYATKINA, K. V. 1956
The Buryats. In: Levin, M. G. – Potapov, L. P. (eds.), *The Peoples of Siberia*, Chicago, University of Chicago Press, 203–242.

WYLIE, A. 1985
The reaction against analogy. In: Schiffer, M. B. (ed.), *Advances in Archaeological Method and Theory*, Vol. 8, Orlando, Academic Press, 63–111.

YABLONSKY, L. T. 1995
Written Sources and the History of Archaeological Studies of the Saka in Central Asia; the Material Culture of the Saka and Historical Reconstruction; and Some Ethnogenetical Hypotheses. In: Davis-Kimball, J., Bashilov, V. A. – Yablonsky, L. T. (eds.), *Nomads of the Eurasian Steppes in the Early Iron Age*, Berkeley, CA, Zinat Press, 201–252.

10

Sophisticated Fire: Understanding Bonfire Pyrotechnologies in Iron Age France

KEVIN ANDREWS

Introduction

Archaeometric investigation of fine-wares from the Auvergne region of France has demonstrated the possibility that the sophisticated decorative repertoire of Iron Age potters depended not solely on the controlled heat energy of kiln structures, but also involved the use of bonfire pyrotechnologies. Methods of examination included scanning electron microscopy with energy dispersive x-ray micro-analysis, X-ray diffraction analysis, refiring experiments and macroscopic examination of decorative surfaces.

The results have allowed the recognition of similarities of production technology for groups of pottery that were assumed to having been produced in very different ways. Prior to the research presented here, it was assumed that sophisticated kilns were employed in the production of all decorated wares with the control of heat energy and firing atmospheres essential to the development of pigments within decorative coatings. The use of kilns was envisaged despite the general lack of archaeological evidence for kiln structures in the study area. By careful selection and pretreatment of clay slips, simple clamp kilns or bonfires were also used, together with the heat energy of smokeless embers which produced conditions for carbonizing designs onto the decorative surface.

This chapter presents an analysis of the results of archaeometric analyses that have allowed a more thorough questioning of the pyrotechnology involved. Consideration is made of the role of bonfires used in pottery production within the context of archaeological models of increased craft specialization. These results challenge our preconceptions of technological 'progress' and focus our attention on the sophisticated use of simpler pyrotechnologies.

Investigation of ancient fire using technologies is an important area of research. Our understanding of the ways in which heat energy was harnessed for production of a wide range of materials has important implications for modelling modes of production. For pottery production, for example, considerations such as raw materials processing, investment in 'capital' installations such as kilns, drying sheds, and supplies of fuel (amongst many others) will all inform our inferences concerning the level of craft specialisation evident in the manufacturing process (*cf.* ANDREWS 1997). Debate has clearly shown that layers of pre-understandings and the ethnocentric conditioning of researchers affect the inferences they make concerning ancient technologies (INGOLD 1988, LEMONNIER 1993, PFAFFENBURGER 1988). A paucity of critical theoretical debate has allowed material remains from past craft activities to be seen as somehow outside the social arena and within its own enclave of 'technology'. Moreover strong, overarching or 'world view' paradigms can colour our interpretations without us really noticing. One such paradigm is that of technological progression. It is assumed for example that through time techniques of manufacture generally become more efficient and sophisticated. Some researchers such as DOBRES and ROBB (2000) are seeking alternative definitions of 'technology' which try to reconnect the web of technical, art, symbol, and social, to allow a more critically informed understanding of ancient technologies.

In this paper the assumption that complex updraft kilns were routinely used in the manufacture of finewares is challenged. In the study area of the Auvergne there is much evidence of craft specialisation (COLLIS 1975a, 1975b, 1980, 1984a, 1984b, 1985, COLLIS *et al.* 1983). Assemblages of fineware pottery reveal a sophisticated repertoire of decorated vessels. There is, however, a lack of evidence of kiln structures in the study area, with a limited number of production sites. Evidence of kiln structures from middle to late La Tène in the Aurvergne has been found at Lesoux, Randan and La Marechal (JOHN COLLIS *pers. comm.*) Previous archaeometric work on Iron Age decorated wares has suggested bonfiring of painted pottery (RIGBY *et al.* 1989), whilst ethnographic observations have shown that sophisticated painted pottery can be successfully produced in open bonfires (COX *et al.* 2000). POOL (2000) demonstrates the point that in some ethnographic and historical cases both bonfiring and kilns are used within the same potting tradition.

This paper focuses on the technology of production of a particular Auvergnian fineware (La Tène Black burnished ware). Results of archaeometric investigations help to underline the importance of the above introductory remarks with respect to modelling the pyrotechnology used in the manufacture of the pottery and the implications of the results in considering concepts of craft specialisation and challenging concepts of inevitable technological progression.

Pottery Types Investigated

All the fineware ceramics investigated are from the study area of the Auvergne in central France (*Figure 10.1*). They date to the period delimitated chronologically as the Iron Age and early Roman (500 BC – 50 AD).

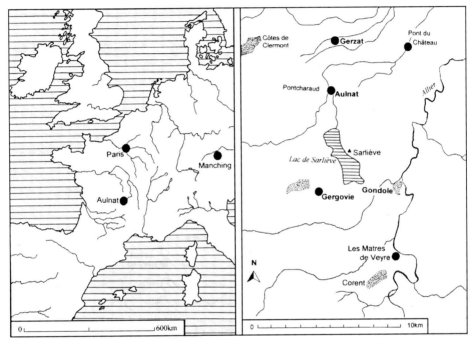

Figure 10.1. Location map showing principal sites of
Aulnat and Gerzat and the Auvergne region.

There is a large range of coated (or suspected coated) wares within the typical excavated assemblage. This range includes the 'painted pottery', black-burnished wares, grey reduced wares, white slipped flagons and a range of less well defined groups such as red-slipped wares and the wares produced by the development of the Samian industries towards the end of the Iron Age period. The later finewares also include samples such as the Campanian wares imported from Italy and imitation Campanian wares which were probably attempts by local potters to reproduce valuable imports using local resources (MOREL 1978), and represent copies of exotic styles using local materials and technology.

The principal fineware group discussed here are the La Tène 'black burnished' wares. In areas such as the Auvergne where the clays are fine and homogeneous, body fabric and inclusion study, although helpful, are unlikely to provide answers to questions relating to the technical relationships between the various fineware ceramic types (CUMBERPATCH 1991). Archaeometric research has to be focused onto the nature of the decorative coatings themselves.

The surface of La Tène 'black burnished' ware may be coated with a black/ grey slip which has then been burnished to produce a uniform, lustrous finish. It is also possible that the burnishing action itself produced the coating by pulling the finer fraction of clay to the surface and pushing coarser inclusions into the body fabric, or pulling them out altogether, to create an appearance of a separate coating layer (RICE 1987, 355). In the Late Iron Age in the Auvergne, a relatively common fineware class known as 'central Gaulish rouletted ware' is typical of the assemblage. This ware has a consistent grey core and surface and is probably a reduced ware. This late ware has been confused with the La Tène black burnished ware that has been shown (ANDREWS 1993) to be distinct from the reduced wares in its production technology.

La Tène Black Burnished Ware (Ltbbw)

La Tène black burnished wares can be defined as native medium to fine textured wares of simple bowl, vase and plate forms that are characterized by dark, burnished surfaces. The material examined as part of this research were collected from two sites within the Auvergne Archaeological Survey, namely Aulnat, and Gerzat-Patural. *Figure 10.2* illustrates some typical forms of bowls and a rare bulbous vase-like vessel. Burnishing was employed as a means of decoration mostly on the exterior of pots in the form of burnishing which produced an attractive luster or, more rarely, selective burnishing to produce geometric patterns. Interior surfaces of bowls and dishes were also often burnished, and here the burnish seems to have been a means of providing a functional surface in that its smoothness imparted a degree of impermeability to the surface, possibly enhancing their function as table ware. The phenomenon of burnishing causing a reduction in permeability is due to the creation of a dense surface of fine compacted particles (HENRICKSON – McDONALD 1983, 633).

La Tène black burnished pots have all been wheelthrown. The degree of burnish varies, but is often extremely well executed. This ware has been recovered from the sites of Aulnat and Gerzat-Patural in large quantities, and is the commonest of the finewares under investigation. Literally hundreds of crates of material have been collected. Paradoxically it is the least studied archaeometrically (or otherwise) and systematic work on this ware is still in progress. Traditionally it has been referred to as either 'grey reduced ware', 'ceramique fumigee' (smoked ware) or simply as 'black burnished ware'. It remains until now undefined. It is referred to here as 'La Tène black burnished ware', to avoid confusion with the well defined Romano-British black-burnished ware described by PEACOCK (1982), and FARRAR (1973).

Figure 10.2. Typical examples of La Tène black burnished ware.

La Tène black burnished ware has been recovered from features representing the early to late chronological phases. It was assumed to be produced by a process of simple burnishing of an unprepared surface to produce a luster, and subsequent kiln-firing in a reducing atmosphere. These ideas about production technology were intuitive assumptions. Preliminary examination of the La Tène black burnished ware type showed that much of the material although having very sharply defined black firing cores in fresh fracture, had a largely oxidised and orange to buff fabric (see *Figure 10.3*). Thinner walled grey-reduced wares (often displaying rouletted decoration) were however reduced having a uniform grey fabric. Much of this material had been bagged with the black burnished wares. Whilst systematic work on the black and grey wares remains to be carried out, the La Tène black burnished ware was selected for further analysis since it appeared on closer examination to have a possible coating rather than a simple burnish. Several observations pointed to a more sophisticated finishing technique than simple burnishing. For example on some sherds, post-depositional erosion had worn away areas of the surface in a way that looked very similar to that of the 'painted pottery' class which has been demonstrated to be coated with pigmented slips (ANDREWS 1997). Small fragments of the burnish had also been noted to chip off the sherds – a definite coating structure is therefore inferred. It is also possible that the outer layers of the ceramic are vitrified and not in fact coated and that this accounts for the different physical properties of the outer layers compared to the body. Archaeometric examination of this ware's decorative surface is necessary to try to see exactly how the colour and lustre were produced.

Several questions must be considered regarding the nature of the surface finish of the La Tène black burnished wares.

1) Was the coating-like surface observed on the La Tène black burnished ware produced by simple burnishing, or was the burnish enhanced by the addition of a clay slip layer?

2) If a slip-burnish is observed, was the slip prepared from separate source clays, or was it refined from the same clays as used for the body fabrics ('self-slipping')?

3) How was the black colour of the decorative surface achieved, were mineral pigments added to the surface to produce the black colour, for example?

4) What implications did burnishing have on the firing technology? A study of exactly how the luster and colour of the decorative coatings of the La Tène black burnished ware were produced could have important implications to our models of the firing methods used in the production of this ware.

MOREL (1982) describes similar wares from the site of Feurs in the Loire territory neighbouring the Auvergne as 'black varnished' or 'black finished' ware. They are included in the site report as imports related to the Campanian wares which were imported from Italy (and have also been found on the site of Aulnat). MOREL (1982, 91) suggests that if the majority of the fragments of black wares come from the Mediterranean as imports, other similar wares that are not truly Campanian might be attributed to local workshops using the imported material as a model. Archaeometric work has been carried out on classical Campanian wares (MAGGETTI *et al.* 1981) and proto- Campanian material (VENDRELL-SAZ *et al.* 1991) which can serve as a comparison with the work undertaken here.

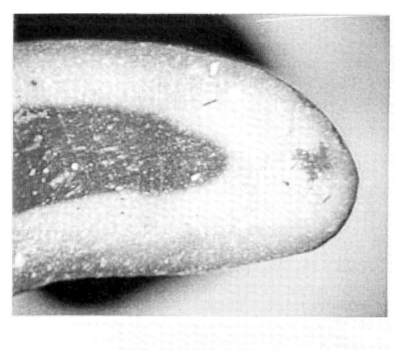

Figure 10.3. Cross-section of La Tène black burnished ware showing distinct firing core within oxidized fabrics.

Analytical Results and their Evaluation

Burnishing is one of the oldest and most primitive techniques of ceramic decoration (NOLL 1982, 152). The burnishing action (rubbing the unfired vessel surface after it has dried, with a smooth, hard tool) often has the effect of producing a thin layer of fine clay particles on the surface as coarser particles are either pushed further into the body fabric, or are pulled out of the vessel altogether as burnishing debris.

Fine clay particles are aligned by the burnishing action to produce a luster owing to the uniformly reflected light off such well-aligned surfaces. Such surface enrichment of clay minerals on the burnished surface can be mistaken for slip layers.

It is important that suspected coatings on burnished wares be carefully examined in order to distinguish between slurried surfaces, intentionally slipped ones, simple burnishes (where the untreated or slurried vessel surface has been burnished) or 'slip -burnishes' where the burnish has been enhanced by a prior application of a refined clay slip.

Irrespective of how the burnish is achieved, such surfaces can be disrupted and lost by over firing the vessel (ABLETT 1974) so that the presence of burnished surfaces on ancient pottery is an important consideration when attempting to model the likely firing conditions. Burnished decoration therefore is worthy of careful examination in the clues that they might hold to our understanding of prehistoric pyrotechnologies.

Such examination should ideally include an evaluation of the structural characteristics of the burnished surface, as well as its chemical composition. Similar chemical compositions between burnished ceramic finishes and underlying body fabrics do not necessarily imply that the enrichment of clay minerals at the surface is a function of the burnishing action. Fine clay fractions can be separated from the same clays as used for the body fabric and re-applied to the surface as a slip suspension. This would enhance subsequent burnishing, as the slip would provide a source of fine clay particles that are aligned by the burnishing action. The application of a slip as a surface finish, which has been prepared from the fine fraction of the same clays as used for the body fabric, has been termed 'self-slipping' (RICE 1987, 151). 'Self slipping' followed by burnishing implies a greater degree of investment into the decoration technology, since it involves more production steps than burnishing alone.

Several analytical techniques were employed to investigate the decorative surfaces of the La Tène black burnished ware. These included: optical microscopy, scanning electron microscopy with associated energy dispersive spectroscopy, and X-ray diffraction studies. To better understand the firing techniques used in the manufacture of these finewares, a series of refiring experiments was also carried out.

Optical Examination and Sem/Qualitative EDS

The first stage of the analysis was simple examination of the untreated sample sherds using a x10–x50 binocular microscope with oblique fibre optic lighting that effectively illuminated such features as burnishing facets and the textural condition of the surface finishes. Certain features were also examined using a video Scopeman.

It was clear from this preliminary examination, that the black wares are a diverse group. It has already been noted that often 'blackness' had been used as a category in sorting the pottery recovered from the sites, so that the burnished ware collections contained rarer samples of imported black Campanian wares, Roman terra nigra, and grey-reduced Gaulish rouletted wares.

The variations noted by preliminary examination of the La Tène black burnished ware group were as follows:

1. The 'classic' LTBBW (mostly sherds from simple bowl forms) characterised by a highly burnished 'thick' looking dense, black surface. The burnishing facets are very fine (often appearing as if executed whilst the pot was spun on some form of rotary device) and often indiscernible, suggesting polishing of the surface. The body fabric is micaceous, finely textured but does contain abundant inclusions (mainly quartz). There are orange to buff oxidized bands within the sherd cross-sections, often with sharply defined black firing cores. There is variation of body fabric colour with some appearing grey (but still containing a darker, well defined firing core).

2. On another group recognized as potentially different the decorative surface appears as a grey wash through which the pinkish body fabric can be seen. The thin grey coating is not generally burnished. On some sherds, however, the thin grey 'wash' gives way, where it appears to be thicker and darker, to a surface which is burnished and is indistinguishable from 1 above.

3. A third group of sherds are not uniformly burnished, but display a pattern burnish of simple band or complex curvilinear designs. The line patterns are

burnished into the black coated surface with what appear to be single strokes. Sherds of this group are, generally but not invariably, composed of rougher body fabrics, and burnished areas are not as finely finished as group 1.

4. An interesting group does appear to be markedly different from the remaining groups in that, whilst the surface is characteristically of the La Tène black burnished ware type, the body fabric is post-depositionally degraded, highly calcitic and poorly tempered. It decomposes readily when tested with dilute Hydrochloric acid. This fabric type is well known as "grot pot" by ceramicists working on the Auvergne material. The forms of 'grot pot' vessels are virtually indistinguishable, as the fabric is usually encountered as amorphous 'lumps' of severely degraded material. Whilst a common fabric type, samples displaying the 'classic' burnished black surface are extremely rare.

5. The fifth 'group' identified represents sherds displaying similar qualities to group 1. It differs only in the quality of the burnish. Unlike group 1, which has a characteristically carefully finished burnish with extremely, fine and closely packed burnishing facets, those of group five have a crudely finished surface with broad and irregular burnishing facets that are easily observed.

6. Finally there is variation in the coatings of the La Tène black burnished ware class in the amount of micaceous particles present in the finishes. There is likely to be a continuum from medium finishes which contain less mica than the final group which has been identified as containing abundant silver coloured mica particles which give this group a metallic sheen. This 'group' has characteristically an extremely smooth and hard surface with no discernible burnishing facets.

It is likely that some of the differences noted above may be the result of the natural variation within a single potting tradition caused by the production processes involved, or natural variation within sources of clay and other materials used to produce the pottery. Variations may also be caused by post-depositional factors. Certainly on several occasions, what were thought to be distinct surface finishes were observed together on larger sherds and in such cases these differences can be attributed to variation within the single class, and/or post depositional variations. For example, if the vessel illustrated by *figure 10.4* is examined, it may be argued to display most of the 'group' qualities as discussed above (with the exception of group 4). The neck and shoulder portion of the vessel could be classified as group one for example, whilst the broad pattern burnish striped portion would yield sherds of group three.

All the groups (except number 4) were examined by qualitative EDS and XRD and revealed no significant variations between them. On archaeometric grounds then at least there is but one class with the noted variations within this class. It is likely that group 4 represents a departure from the LTBBW and therefore a distinct type. No archaeometric examination was carried out on the small amount of group 4 material available owing to its rarity and potential for further work, whilst archaeometric examination concentrated on group 1.

Figure 10.4. La Tène black burnished ware vase. Most fragmented remains are sherds of simple bowl forms. This bulbous vase would appear to be a rarer form. The luster and quality of the burnish is most evident on the neck and shoulder of the vessel. The remainder of the vessel is only selectively burnished. The geometry of the burnished lines would suggest that the burnishing were performed on a wheel.

Further optical examination of the sherds and their surface finishes was carried out using thin section samples viewed with transmitted light, and epoxy resin mounted samples which had been diamond polished and were examined using reflected light. A Nikon Optifot microscope was used for such examination.

Examination of representative samples of La Tène black burnished ware prepared in thin section reveal that there is enrichment of black colouration at the surface. *Figure 10.5* shows a typical profile through a thin-sectioned sample of La Tène black burnished ware. There is a dense black colouration at the surface that grades into the lighter buff colour of the body fabric the surface of this sherd was smooth and burnished. Such black layers may mask the identification of slip layers, which are generally poorly preserved in thin section preparations. Importantly, however, the profile of dark colouration is revealed. As will be discussed later, the cause of this colouration is likely to be amorphous carbon. Spot EDS analyses of such profiles (of similar examples prepared as polished sections mounted in epoxy resin blocks) from the surface to the interior of the body fabric showed no appreciable increase in chromatophonic elements such as iron in the surface when compared to the body fabric.

Figure 10.5. Micrograph showing the carbon profile through a thin section of La Tène black burnished ware.

Figure 10.6 also illustrates the surface treatment of simple burnishing with no application of a slip prior to the burnishing action. There is little evidence for a surface layer created by the action of burnishing. This particular example, although displaying burnishing facets and a lustrous surface, was not as well finished as the surface illustrated by *figure 10.7* that shows a slip-burnish structure.

Figure 10.7 shows a clear slip structure. This appears to be a deliberate slip layer of fine clay and mica particles applied prior to burnishing since it overlays voids and large quartz particles from which no fine clay particles could possibly have been drawn. Also since the thickness of a coating produced by burnishing is relative (amongst other things) to the pressure applied in the burnishing action, one would expect a burnish coating generated by simple burnishing to be relatively thin, and of uneven thickness. This slip-burnish layer is in the order of

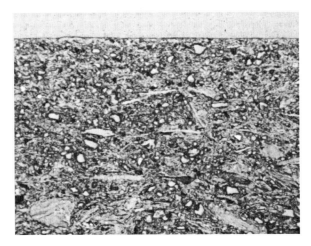

Figure 10.6. Micrograph of cross-sectioned La Tène black burnished ware sample illustrating simple burnishing with no application of a slip prior to the burnishing action. The luster on this sample was not as even as that present on a slip-burnished sample. This figure should be compared to figure 10.7 which shows a slip-burnish structure.

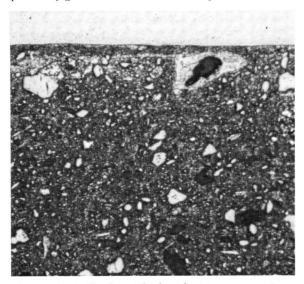

Figure 10.7. Slip-burnished surface in cross-section.

40 microns thick and is uniform, very similar in structure to the slips found on the slip-decorated pottery.

Several features of the La Tène black burnished ware class pointed to the use of slip-coatings in addition to the burnished finish although the slip-burnish is not present on all the sherds examined. The feature most commonly observed which suggests a coating structure is that of surface cracking, and the fact that burnished finishes present on the ware were seen on some samples to spall off in a manner which suggests a coating structure applied to the body fabric.

Many of the pieces on first examination, apart from variation in quality of the burnished finishes, did not appear to display any other decorative techniques than burnishing or slip-burnishing and variations of pattern burnishes. Closer examination of the sherds using the Scopeman system displayed other decorative elements which had been not previously been noted. Rarely bands had been incised into the burnished surface, some of which contained a white material. Testing the white material adherent to the scratched bands with a micro- pipette containing dilute HCl showed that it was composed of a calcitic based pigment, which probably represents a post-fire application of white pigment. ELLIS (1984, 119) reports on the use of similar decorative techniques for the Cucut eni-Tripolye Early Bronze Age cultures of Ukraine where the use of white powder (calcium carbonate) to fill spaces produced by excising (as well as application of heamatite, both post-firing techniques) is in evidence. A simple replication experiment was carried out to reproduce bands by scratching away the burnished surface with a piece of glass. It would seem that such a simple technique for producing a light coloured band would be effective in that the burnish is selectively removed to produce a lighter, matte contrast.

Whilst rare, post-firing painting using red pigments is also observable on the La Tène black burnished ware. The red pigment, showing vestiges of brush marks, was probably held together by an organic binder. Likely candidates for the origin of the pigment are red ochres (ferrous oxides). Such pigments are fugitive and rarely survive burial. More permanent red colours can be produced by the formation of red ferrous oxides such as haematite during the firing (under oxygen rich conditions) of clay slips. That the red pigment is a post-fire application can be inferred by the fact that the black burnished surface is coloured most likely by amorphous carbon or reduced iron oxides, both of which require a smoky non-oxidising atmosphere (see later). It is notoriously difficult to obtain black and red fired pigments on the same vessel, since the red fired pigment based on ferrous oxides, require an oxygen rich firing atmosphere. Manganese based black pigments are the only ones which can be fired together with red fired pigments in an oxidising firing (NOLL *et al.* 1975). The qualitative EDS results showed that manganese based pigments are not employed in the production of the La Tène black burnished wares.

Examination of the surface under the scanning electron microscope shows the closely packed striations of the burnishing facets that would suggest that the burnishing was carried out on some sort of wheel (see *figure 10.8*). The surface is extremely smooth, as can be seen by *figure 10.9* which shows the closely packed clay and mica particles aligned to the surface.

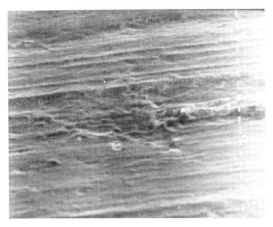

Figure 10.8. SEM micrograph showing the striated nature of the burnished surface of La Tène black burnished ware (scale bar divisions = 10 micrometers).

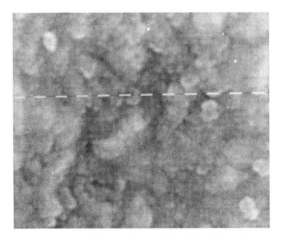

Figure 10.9. SEM micrograph of the surface of La Tène black burnished ware sample (scale bar divisions = one micrometer). Individual clay and fine mica particles can be seen aligned parallel to the surface appearing as oval plates c. 2 micrometers in diameter.

Figure 10.10 shows that there is a layer of fine particles at the surface. The relationship between this layer and the body fabric in fresh fracture as seen here is not as clear as with polished sections illustrated by the optical micrographs. Other samples (not illustrated) of grinding debris show very similar slip-burnish finishes. Small flakes of the slip-burnish were separated from the body fabric during preparation of the samples for refiring experiments. *Figure 10.10* shows that the fine clay particles that make up the slip layer are not fused to any degree consistent with a fairly low firing temperature.

Figure 10.10 shows the open porous structure of the body fabric clay phases. The platey nature of the clay particles have maintained their particle morphology having been little modified by the heat treatment of the firing process which again would suggest that the firing temperatures achieved in the firing of this ware were quite low.

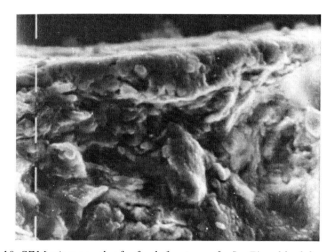

Figure 10.10. SEM micrograph of a fresh fracture of a La Tène black burnished ware sample (scale bar divisions = 10 micrometers) showing the edge of a burnished surface. Unsintered clay particles can be seen within the surface layer.

Qualitative energy dispersive spectroscopic determinations of the chemical composition of areas selected under the scanning electron microscope could be made of representative samples coated with carbon. EDS results taken from area analysis of the surface slip-burnish cross-section shows that the coating is clay based (as evidenced by strong peaks for aluminium and silicon) and probably a calcareous clay with calcium (Ca) being recorded in the spectra. Relatively high levels of potassium (K) could be indicative of a high potash content of the clay (likely, as we shall see from X-ray diffraction results, to be illitic), or may represent an ash component which was added to a slip preparation to act as a deflocculant in settling out the finer fraction of the body clay to be used in the slip burnish. Equally the source of the potassium peak in the spectra could be derived from the mica (hydrous silicates of aluminium, potassium etc.) which have already been noted from the optical examination of the La Tène black burnished ware finish.

Iron is the only other major element to be recorded by the EDS qualitative analyses as a typical component of the slip-burnish. Reduced iron has been noted as a major ceramic pigment giving a black colour, however, as the refiring experiments show (and as will be discussed later) it is unlikely to be the major contributor to the black colour of the La Tène black burnished wares. The iron peak probably derives from natural iron contained in the ferruginous clays used to prepare the slip-burnish coating.

The coatings contain more aluminium, compared to the body fabric. The content of aluminium can be used as an index of the amount of fine clay particles (FREESTONE 1982, 110). Such a result is consistent with the hypothesis that the slip provides a source of fine-clay particles to enhance the subsequent burnishing.

Semi-Quantitative EDS

The epoxy-resin mounted, diamond polished samples examined optically using the Nikon microscope, were also suitable (after sputtering with carbon) for semi-quantitative analysis of their chemical composition using an energy dispersive spectrometer attached to the SEM.

Importantly the semi-quantitative EDS results confirm that manganese is not an important element in this ware, and is unlikely to play a chromatophonic role. Similarly the suggestion that iron sulphides could be responsible for black colouration of ceramic surfaces (HOFMANN 1966, 218) can be discounted for the La Tène black burnished ware since the percentage sulphur detected in the coatings is minimal (a mean of less than 2%) and if a small amount of iron sulphide is present, it is unlikely to be making a significant contribution to the black colouration.

An important feature of the semi-quantitative analysis is that it has shown that calcium is present in sufficient quantity to classify the La Tène black burnished ware as calcareous (over 5% calcium). Is there then addition of calcitic inclusions, or deliberate selection of calcareous clays for La Tène black burnished wares? Calcium acts as a fluxing agent enabling sufficient body strength even at relatively low temperatures, which is an important consideration for wares fired in a bon-fire or open firing.

X-Ray Diffraction

Samples of slip-burnish surface finishes and respective body fabrics were taken from representative sherds and subjected to X-ray diffraction analysis using a Phillips PW1373 diffractometer. In total 20 diffractograms were taken from samples of sherds displaying a slip-burnished finish. The results were repetitive. A typical diffractogram shows that the results for the slip burnish and the body fabric are broadly similar implying that they are of the same bulk composition. That is, the XRD results comply with the qualitative and semi-quantitative EDS results in suggesting that the slip burnish was created by self-slipping whereby the slip was refined from the body clays.

The diffractograms recorded for the La Tène black burnished ware are uniform in the representation of illitic or micaceous clays with the peaks for Illite/mica being the most notable features of the diffractograms.

Predominance of illitic/micaceous clays offers advantages of a platey particle morphology capable of producing a smooth surface. It is generally agreed that an illitic or micaceous clay mineral should predominate in a slip clay (e.g., FREESTONE 1987, 110) (illite is a clay mineral with a structure like that of mica).

Illitic clays are naturally rich in potash an effective flux. Iron oxides and hydroxides in feruginous clays or added in the form of ocherous earths will under reducing conditions form spinel phases which are black – the most common pigment mineral so formed is hercynite ($FeO.Al_2O_3$) and its solid solution series and magnetite (Fe_3O_4) (STOUT – HURST 1985, 29). Iron rich silicates containing a high percentage of ferrous ions can also occur naturally in some clays and are another possible cause of black colouration, provided they are not oxidised during firing. All these minerals lose their crystallinity above 700° C that renders them invisible to X-ray diffraction detection. Since illite (which is also dehydroxylised at around 700° C) is present, hercynite or magnetite, if present, should also be detectable. However, they are absent as were other well known minerals capable of imparting black colour to ceramic surfaces such as jacobsite and graphite. A lack of vitrification structures or glassy phases in the La Tène black burnished ware would also rule out the possibility of entrapped black ferrous ions within such phases contributing to the black colour. Graphite, another common mineral pigment imparting a black colour to ceramics, was also searched for but found to be absent in the X-ray diffraction analysis of the La Tène black burnished wares. It is possible that such mineral species are present, but have not been recorded by the X-ray diffraction analysis owing to their fine-grained nature. Any mineral-based pigments that reside in very small grains will not be detected. As such the involvement of such pigments cannot be conclusively ruled out. Based on our knowledge of ceramic pigments however, we can suggest that amorphous carbon is likely the major cause of colour (a conclusion further supported by the results of the refiring experiments – see later).

The mineral calcite is also represented by the X-ray diffraction results. The presence of calcite accounts for the relatively high values of calcium noted in the results of the EDS analyses. The decomposition of calcite in mixtures with clay minerals begins at *c.* 650-700° C but the bulk of the calcite disappears at about 800° C (HEINMANN 1982, 90) so that firing must have been below this temperature for the calcite to have retained its crystallinity and therefore to have been susceptible to analyses via X-ray diffraction studies. At the other end of the temperature scale, PETERS and JENNI (1973) report that diopside mineral is produced in calcareous clays at temperatures above 850° C. If this temperature is not achieved in the firing, the resultant ceramic contains not diopside, but unaltered calcite. The fact that diopside is absent from the X-ray diffraction results, is consistent with the low-firing temperature hypothesis for the La Tène black burnished wares. Again this negative evidence agrees with results of the re-firing experiments.

The results of the X-ray diffraction analyses discussed above have pointed to the use of illitic clays as the basic ingredient for the slip involved in the slip-burnish technique used in the characteristic finish of the La Tène black burnished ware. The burnishing of an illitic slip applied to the surface of the leather hard product (the illitic slip providing a source of fine particles which would form a high burnish undisrupted by inclusions which are common in the body fabric), would enable the potter to produce a higher quality finish than would be afforded by the burnishing of the untreated vessel surface.

Refiring Experiments

The refiring experiments involving the heating of sample sherds to successively higher temperatures in a programmable kiln allowed observations to be made concerning changes to surface colour and luster of the samples. Onset of linear shrinkage detectable through stepwise measurement of sample dimensions also allowed maximum firing temperatures to be inferred. The estimated firing temperatures for La Tène black burnished ware were uniformly low with 64% of the sample displaying estimated firing temperatures below 850° C and none above 875° C. In comparing these results with the ethnothermometric data presented in GOSSELAIN's (1992) paper, it can be suggested that this restricted temperature range is characteristic of a pit firing in which insulating sherds have been used to separate the pots from the fuel. It would be rash to conclude so specifically that this was the firing regime used in the production of the La Tène black burnished ware since the restricted range is probably owing to the fact that there were only four references available to Gosselain for this firing structure. There are however thirty references (GOSSELAIN 1992, 246) to support the restricted range of the open firing using insulating sherds and the open pit firing. To infer that the La Tène black burnished ware was fired in either of the three structures displaying such a restricted range (i.e., open bonfire with insulating sherds, open pit, or pit firing with insulating sherds) would probably be a safer inference. It could further be argued however, that to maintain the burnished surface, the La Tène black burnished ware would require insulation from the fuel (*cf.* SARGENT 1983, 52), so that the open pit firing could also be ruled out, leaving us with two strong possibilities.

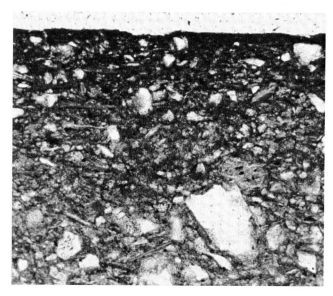

Figure 10.11. Thin section of La Tène black burnished ware
showing abundant quartz inclusions.

It is also important to mention that the shape of the pot helps to impart strength to low-fired ceramics, for example bowl and plate forms. It is noticeable that *c.* 80% of the La Tène black burnished ware sample consists of such simple forms, as far as can be ascertained from the fragmented remains. A possible objection to the scenario of the bonfiring or clamp-roasting of the La Tène black burnished ware is the fact that it is a fine-ware and as such would not be able to withstand the stresses involved in an open firing. It is maintained that it is impossible to bonfire fire a vessel without the fabric being 'crammed' full of filler in order for gasses and water vapour to escape (NICKLIN 1979, 456; RYE 1976, 109). With careful drying of the vessels and the use of abundant quartz inclusions (as has been illustrated by *figure 10.11*), however and the noted use of highly micaceous clay would enable such wares to be fired successfully in simple clamp kilns. ELLIS (1984, 114) has suggested that micaceous clays need little added inclusions/temper and that a high mica content are sufficient to prevent the extensive clay shrinkage and subsequent cracking.

The La Tène black burnished wares lost their colouration after the first firing to 500° C. A set of further experiments were conducted to ascertain more closely at what temperature black colouration is lost. Samples were refired at 50° C intervals from 100–500° C. The black coatings (usually overlying oxidised fabrics) had all lost their black colour by the firing 400–450° C, although this phenomenon had began as low as 250° C pointing to amorphous carbon (which is burned out at these fairly low temperatures) as the main contributor to the black colour. Similar conclusions have been drawn based on the loss of black colour at low temperatures on Iron Age finewares from the Heuneburg (MAGGETTI – SCHWAB 1982, 30) and is a noted phenomenon and diagnostic of amorphous carbon coloured surfaces (NOLL 1976). This colour change is illustrated by *figure 10.12*. The carbon was probably introduced by smoking the ware by heating the finished ceramic in a soot-laden atmosphere. Any secondary burning in an oxidising atmosphere would have the effect of burning out the carbon based colour, which may explain the 'negative fireclouding' features which are fairly common for this class of pottery (see *figure 10.13*). That is areas of the surface in which the black burnish surface, whilst still retaining the burnish, had lost its black colouration and taken on the colour (pale buffs) of the body fabric. Fireclouding is the opposite of this whereby lighter coloured vessel surfaces are clouded by darker coloured soots from the firing process or secondary burning (MAGGETTI – SCHWAB 1982, 33).

Consistently the burnish or polish of the sherds was destroyed at temperatures within the region 725–750° C. Experimental work by ABLETT (1974), a craft potter

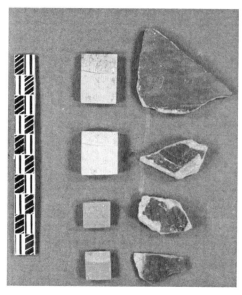

Figure 10.12. Refiring experiments showing loss of black colouration at low temperatures through comparison of refired sherds and control (un-refired) sherds.

Figure 10.13. La Tène black burnished ware bowl showing negative fireclouding.

interested in burnishing, has shown that in order to retain the luster of a burnished surface, the firing has to be quite low so that there is minimum disturbance to the aligned clay particles which impart reflected luster to the finish. Any firing above 700° C tends to dull the surface of the burnish (ABLETT 1974, 3). The presence of calcium (as revealed by the EDS and XRD results) would have acted as an effective flux, lowering the onset of the liquid phase formation for the La Tène black burnished wares. The actual firing temperature is likely to have been lower, as revealed by the loss of the burnish at lower temperatures.

Discussion

Several questions were posed concerning the nature of the decorative surfaces present on the La Tène black burnished wares. To what extent have these questions been answered by the results obtained from the various analyses described above?

The EDS analyses, together with the XRD results have shown that the slips used in the slip-burnish technique were derived from the same source clays as was used for the body fabric in a process known as 'self-slipping'. The use of slips prepared from the same source clays as used for the vessels has the advantage of having a single supply for both materials. It also has the technological advantage in that the adherence of coatings prepared from body fabric clays have a similar firing behaviour and are therefore fired in sympathy with the vessel fabric leading to good adherence.

The question of how the black colouration of the ware was achieved has been answered largely by the XRD and refiring experiments. Scanning electron microscopy has also shown that there is no developed glassy matrix which could acts as a medium for reduced iron colouration (*cf.* HEDGES 1975). Whilst iron has been shown to be present, lack of iron based mineral pigments in the XRD results, together with the observed colour loss recorded by the refiring experiments lead to the conclusion that amorphous carbon is the cause of the black colouration. There are several ways in which carbon could have been introduced to the burnished slip. Firstly, the fine fraction of the source clay for the self slip may have naturally contained high organic component. Secondly, carbonaceous organic materials such as oils or vegetable extracts may have been added to the slip preparation. Finally, the slip-burnish may have become impregnated with carbon during firing in a reducing atmosphere, or as part of a post-firing smoking process. The thin section results showing the gradation of black colouration into the surface would support the last of these possibilities, suggesting that the carbon was diffused into the ware by some form of smoking process. Such a technique has been observed ethnographically on similar wares from Norway, and has been postulated as the decorative technology employed on Iron Age burnished wares in Denmark (STOUT – HURST 1985, 225).

The need to provide a uniform colour for the La Tène black burnished ware by smoking in a soot-laden atmosphere ties in with the conclusions formed concerning the firing technology employed in the production of this ware. Simple clamp kilns or open bonfirings produce wares which often have a mottled surface colour owing to the nature of the firing method, the firing atmospheres being notoriously difficult to control. The use of the smoking technique to uniformly colour the ware would effectively mask any mottling produced by the firing process, as well as providing an aesthetically pleasing finish. Infusing the surface with amorphous carbon also has the effect of enhancing lustrous surfaces (HEGDES 1979, 142).

In conclusion, it is probable that the La Tène black burnished ware were often produced with the use of the slip-burnishing technique, although the nature of the micaceous clays allowed for a reasonable burnished finish on untreated vessel surfaces. For those vessels requiring a higher quality burnish however, the slip-burnish was employed. The wares were bonfired or roasted in clamp kilns at temperatures that would not overly disrupt the luster of the surface finish, rather than highly fired in archaeologically robust up draft kilns. Thus the absence of kiln structures in the area does not have to be explained away as excavation bias. The analytical results have shown that black pigments involving reduced iron oxides (a common means of producing a black colour in prehistory – especially in the Mediterranean) was not a feature of the production of La Tène black burnished wares. The surfaces were coloured by some kind of smoking process in which amorphous carbon was introduced into the surface layers.

The results of the archaeometric analyses presented here have demonstrated that there are recognisable similarities in the mode of production of what first appear to be very different potting styles. When underlying assumptions of craft specialisation and progress are challenged through a critical reappraisal of how we understand and interpret the material remains of past technology, we open up the possibility of seeing those technologies in different ways.

Acknowledgements

The author would like to thank Gilly Cox and Alan Watchman for discussions concerning their ethnographic observations of pottery bonfiring. John Collis is also acknowledged for allowing materials from the Auvergne Archaeological Survey to be studied.

Bibliography

ABLETT, J. 1974
Burnishing. *Pottery Quarterly* 11 (41), 2–5.

ANDREWS, K. 1993
Slip on something luxurious: The technology of Iron Age fineware pottery production – an archaeometric study of ceramic decorative finishes on pottery from the Auvergne region of France. Unpublished PhD thesis. University of Sheffield.

ANDREWS, K. 1997
From ceramic finishes to modes of production: Iron Age finewares from central France. In: Cumberpatch, C. – Blinkhorn, P. (eds.)*, Not so much a pot, more a way of life*, Oxbow monograph 83, 57–75

COLLIS, J. R. 1975A
Defended sites of the Late La Tène. Oxford: BAR. International Series 2.

COLLIS, J. R. 1975B
Excavations at Aulnat, Clermont-Ferrand: a preliminary report with some notes on the earliest towns in France. *The Archaeological Journal* 132, 1–15.

COLLIS, J. R. 1980
Aulnat and urbanisation in France: a second interim report. *The Archaeological Journal* 137, 1–15.

COLLIS, J. R. 1984A
Oppida. Earliest towns North of the Alps. Sheffield University

COLLIS, J. R. 1984B
Aulnat and Urbanisation: the theoretical problems. *Études Celtiques* 21, 111–117.

COLLIS, J. R. 1985
The European Iron Age. Batsford Books.

COLLIS, J. R. – DUVAL, A. – PERICHON, R. 1983 (eds.)
Le Deuxieme Age du Fer en Auvergne et en Forez et ses relation avec les regions voisines. Sheffield University.

COX, G. – WATCHMAN, A. – BLAKE, K. 2000
Paint and slip compositions of Papua New Guinea pottery, *32nd International Symposium on Archaeometry Program Abstracts*, Mexico City: Instituto de Investigaciones Antropologicas, 226–227.

CUMBERPATCH, C. G. 1991
The Production and Circulation of Late Iron Age Slip Decorated Pottery in Central Europe. University of Sheffield unpublished PhD. thesis.

DOBRES, M. – ROBB, J. 2000
Agency in Archaeology, London, Routledge.

ELLIS, L. 1984
The Cucuteni-Tripolye culture. A study in technology and the origins of complex society. Oxford: BAR. International Series 217.

FARRAR, R. A. H. 1973
The Techniques and sources of Romano-British black-burnished ware. In: Detsicas, A. P. (ed.), *Current research in Romano-British pottery.* London 67–103.

FREESTONE, I. C. 1982
Applications and potential of Electron Probe micro-analysis in technological and provenance investigations of ancient ceramics. *Archaeometry* 24(2), 99–116.

FREESTONE, I. 1987
Ceramic Analysis. In: Mellars (ed.), *Research Priorities in Archaeological Science*, CBA Report.

GOSSELAIN, O. P. 1992
Bonfire of the Enquiries. Pottery Firing Temperatures in Archaeology: What For? *Journal of Archaeological Science* 19, 243–259.

HEDGES, R. E. M. 1975
Mossbauer Spectroscopy of Chinese glazed ceramics. *Nature* 254, 501–503.

HEGDES, K. T. M. 1979
Analyses of ancient Indian deluxe wares. *Archaeo-physika* 10, 141–55.

HENRICKSON, E. F. – MCDONALD, M. M. 1983
Ceramic form and Function: an ethnographic search and archaeological application. *American Anthropologist* 85, 630–643.

HOFMANN, U. 1966
Die Chemie der Antiken Keramik. *Naturwissenschaften* 53, 218–23.

INGOLD, T. 1988
Tools, mind and machines: an excursion in the philosophy of technology. *Techniques at Culture* 12, 151–61.

LEMONNIER, P. 1993
Technological Choices: Transformation in Material Cultures since the Neolithic. London, Routledge.

MAGGETTI, M. – GALETTI, G. – SCHWANDER, H. – PICON, M. – WERSIKON, R. 1981
Campanian pottery, the nature of the black coating. *Archaeometry* 23(2), 199–207.

MAGGETTI, M. – SCHWAB, H. 1982
Iron Age Fine Pottery from Chatillon-s-Glane and the Heuneburg. *Archaeometry* 24(1), 21–35.

MOREL, J. P. 1978
A propos des céramiques campaniennes de France et d'Espagne. *Archéologie en Languedoc* 1, 149–168.

MOREL, J. P. 1982
La Céramique à vernis noir de Carthage-Bursa: nouvelles données et éléments de comparaison. In: *Actes du colloque sur la céramique antique, Carthage CEPAC*, 43–76.

NICKLIN, K. 1979
The location of pottery manufacture. *Man* 14, 436–458.

NOLL, W. 1976
Kaltbemalung antiker Gefässkeramik. *Naturwissenschaften* 63, 384.

NOLL, W. 1982
Mineralogie und Technik der Keramiken Altkretas (Mineralogy and Technique of the ceramics of ancient Crete). *Neues Jahrbuch für Mineralogie Abhandlungen* 143(2), 150–199.

NOLL, W. – BORN, L. – HOLM, R. 1975
Keramiken und Wandmalereien der Ausgrabungen vom Thera. *Naturwissenschaften* 62, 87–94.

PEACOCK, D. P. S. 1982
Pottery in the Roman World. London, Longman.

PETERS T. S. – JENNI, J. P. 1973
Mineralogische Untersuchungen über das Brennverhalten von Ziegeltonen. *Beitr. zur Geologie der Schweiz, Geotechnische Ser., Lief.* 50, 1–59.

PFAFFENBERGER, B. 1988
Fetished objects and humanised nature – towards an anthropology of technology. *Man* 23, 236–252.

PFAFFENBERGER, B. 1993
Social anthropology of technology *Annual Review of Anthropology* 21, 491–516.

POOL, C. A. 2000
Why a kiln? Firing technology in the Sierra de Los Tuxtlas, Veracruz (Mexico). *Archaeometry* 41(1), 61–77.

RICE, P. M. 1987
Pottery Analysis. University of Chicago Press.

RIGBY, V. – MIDDLETON, A. P. – FREESTONE, I. C. 1989
The Prunay workshop: technical examination of La Tène bichrome painted pottery from Champagne. *World Archaeology* 21(1), 1–16.

RYE, O. S. 1976
Keeping your temper under control: materials and the manufacture of Papuan pottery. *Archaeology and Physical Anthropology in Oceana* 11(2), 106–137.

SARGENT, P. 1983
Wood-the creative element. *Pottery Quarterly* 14(54), 51–74.

STOUT, A. M. – HURST, A. 1985
X-ray diffraction of Early Iron Age pottery from Western Norway. *Archaeometry* 27(2), 225–230.

VENDRELL-SAZ, M. – PRADELL, T. – MOLERA, J. – ACIAGA, S. 1991
Proto-campanian and A-Campanian ceramics: characterisation of the differences between the black coatings. *Archaeometry* 33(1), 109–117.

11

Detecting Ancient Fires and Simple Fireplaces in the Old World

RALPH ROWLETT

Introduction

The detection of the first use of fire by hominids becomes essential for understanding human biological and cultural evolution. WRANGHAM and his co-
workers find that fire control and cooking are critical developments enabling
the emergence of distinctive human behaviors and social organization. Brace
theorizes that early humans could not control fires until they had reached a stage
of cultural development when cultural differences between bands had emerged.
Less basically, it is desirable to detect fires and burning not only for the production
of ceramics, but also for the heat treatment of flint, the use of stone boilers, and
the construction of sweat lodges as well as for a myriad of other purposes, such
as rituals, fire-setting in mining and the use of fire for environmental management
or hunting. Not all of these uses necessarily leave behind a visible concentration
of charcoal and ash, which may be obliterated by various forms of erosion.
Humanly instigated fires could be confused with lightning strikes, forest fires,
grass fires, spontaneous combustion, oxidation, the burning of individual trees
and even fungus. A number of methods and techniques for recognizing these
phenomena archaeologically, and for making positive identifications of hearths
and fireplaces have been developed. This work has helped confirm the presence
of wood fires at Zhoukoudian, which would have facilitated survival at the height
of the antepenultimate glaciation. The oldest fires that we have detected are in
Africa, primarily at Koobi Fora, Kenya, in the Okote Tuff that produced the
Karari Industry and an early form of *Homo erectus* (*H. ergaster*). These fires may
have been intended for predator intimidation, making feasible the base camps
important to the social theory of Rolland.

Burning Desires

Many palaeoanthropologists had been reluctant to believe that even early *Homo erectus*, with a brain *ca.* 900 cu. cm, 29% bigger than the brain of *Homo habilis*, would dare to use and manipulate fire. So the team of Charles Peters and I, ultimately joined by Michael Davis and Robert Graber, were engaged by the late Glynn Isaac, Richard Leakey, Jack Harris, and Harry Merrick, the discoverers of these reddish patches, to determine just what caused them. In the meantime M. Barbetti, Randy Bellomo, and William Kean made independent tests of these patches utililizing archaeomagnetism. We tested various possibilities – that the reddish patches were caused by an African fungus, that they were iron deposits, that they were the results of strikes by the lightning so frequent in Africa, that they were the results of forest and grass fires, or that they were results of individual burnt trees.

Since the early discovery of *Homo erectus* and associated artifacts at Locality 1 of Zhoukoudian, there have been the tantalizing occurrence of *Homo erectus* and evidence of fires and at several sites in Eurasia and, above all, in Africa. Some of these sites include the baked burnt clay at Chesowanja in Kenya and Isernia in Italy, apparent hearths at Escale Cave, Gadeb, and burnt bone at Swartkrans (SILLEN – BRAIN 1990). Although the dates on these sites range from 1.5 to as late 400,000 years ago for uncontestable hearths at Verteszollos, Hungary, none of these other claims for fire have gone uncontested (LOCKE 1999). Even at Zhoukoudian there has been expressed some doubt of burning, most palaeoanthorpologists admit the presence of massive burning at Zhoukoudian, but question whether there was actual control of the fire.

Traces of 1.6 million year delimited reddened patches in the sediments of the Okote Tuff at the classic fossil human site of Koobi Fora on the east shore of Lake Turkana in Kenya would present the best evidence of fire control if the redding was actually caused by burning. Koobi Fora is famous for its early human *Homo habilis* and Oldawan stone tool remains in the KBS volcanic tuff dated two million years ago, but the stratigraphically superior Okote Tuff contains highly significance remains as well. In these Okote Tuffs, potassium argon dated to 1.6 million years ago, have been recovered an early form of erectus, sometimes called *H. ergaster*. These human fossils occur alongside a late form of the robust fossil australopithecine, *Australopithecus boisei*, the one that L.B.S. Leakey called "Nutcracker Man". The stone tools left behind by early *Homo erectus/H. ergaster* also show advances, with distinct forms, such as the evenly denticulated heavy scraper, so that this tool assemblage gets its own designation as the Industry. What is most important about the Okote Tuff at Koobi Fora is the association of these human palaeontological and archaeological materials with reddish spots of ca. 50 cm. diameter, potential candidates for being among the oldest known fireplaces in the world.

The phytoliths found in the Koobi Fora lensatic basins do indeed show this heterogeneity of phytoliths even more than our experimental fireplaces, which had by college students who were "blindly" unaware of the implications of the kinds of fireplaces and stump burnings that they were instructed to make. Even the fireplaces with a single kind of wood fuel tend to have more mixed in phytoliths than the remains of a standing tree (or stump) burnt in place. The pragmatic, unselfconscious fireplace particularly resembled the ancient ones in its phytolith heterogeneity. Of the four red patches initially found by Glynn Isaac, Richard Leakey, and J.W.K. Harris at locality FxJj 20 East at Koobi Fora (ISAAC – ISAAC 1997, fig. 4.29, pl. 4.14), three round red patches can be inferred as fireplaces, while an irregular narrow red patch, at the edge of their excavation, does indeed seem to be a burnt tree for the phytoliths are mostly of the same kind there.

The presence of a burnt tree is truly fortunate, because it helps provide a clarifying comparative perspective. Note how the fireplaces are displayed in an arc, apparently of a circular encampment. In the now recognized fireplaces, the augmented amount of non-grass phytoliths, as compared to the ambient sediments and the burnt irregular patch, would seem to come from the use of the plants as tinder to help start the fires and perhaps also from plant foods prepared in some way with the fires (*Figure 11.1*).

We used the technique Differential Thermal Analysis (DTA) to discover that the fires that caused the reddish patches burned quite well, yet without being inordinately hot – say, as lightning or an iron smelting fire would be attaining temperatures just under 400° C. That is the temperature of a roaring good campfire, without using any special techniques, such as a fan or bellows, to make it burn hotter.

BELLOMO and KEAN'S (1997) archaeomagnetic analyses also concluded that the fires did not burn over 400° C. What is more, palaeoanthropologists R. V. Bellomo and William Kean, in studies totally independent from ours and using archaeomagnetism, confirmed our results that the reddish spots at FxJj M were caused by burning, and provided the added evidence that the fires had burned in more than one instance and had to be re-kindled somehow. In other words, the fireplaces were where people had returned from time to time to re-ignite their fires (BELLOMO 1990; BELLOMO – KEAN 1997). Bellomo and Kean investigated a different series of reddish patches in the locality FxjM and one patch in Fxj20E, the same patch as our reddish spot Nr. 4. They report burning for the patches in FxjM (BELLOMO – KEAN 1997, 223–36) but also report that our reddish spot Nr. 4 is anomalous in its archaeomagnetic responses. Anomalous for us, too, we suggest that its eccentricity result from its being a burnt tree, with massive amounts of the same arboreal phytoliths. Note also that reddish spot Nr. 4 has a more irregular outline than the other patches investigated by Bellomo and his colleagues and by my colleagues and me.

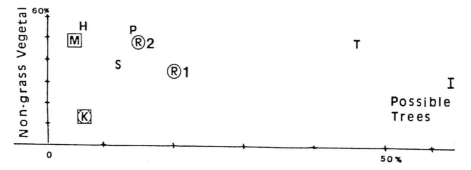

Figure 11.1. Plot of the phytolith category components of the actualistic burnt features, the experimental campfires, and the archaeological features at Koobi Fora FXJj 20East in the Okote Karari sediments. Since the quantities are expressed as percentages, the two dimensional graph incorporates three-dimensional data. Note that the two Koobi Fora Red Patches match well the Karari sediments, just as the experimental and actualistic fireplaces group closely with their soil matrices. The experimentally burnt tree (T) and the irregular reddish patch from Koobi Fora (I) both veer toward the edge of the graph. T = tree base, P = pragmatic campfire, S = single wood species fuel campfire, H = haphazard fuel campfire, M = Missouri soil, K = Karari sediments, R1 and R2 = Koobi Fora "Red Patches" F1 and F2, and I = Irregular Koobi Fora "Red Patch" 3.

The several independent studies of these sharply delimited reddish features at Koobi Fora all point to the same conclusions. That these were the relatively confined fireplaces composed of easily ignited wood that burnt only to normal campfire temperatures. The possibility that these fireplaces were built by later hominids that came from the Karari Industry occupation by *Homo erectus/H. ergaster* was tested as well. While no fossils bones were found inside the fireplaces (unlike the case for stone tools), some bones were found very near the fireplaces as well as away from the fireplaces further out in the sites. The mineralized bones near the fireplaces, if they had been heated after their initial deposition, would have shown a reduced ESR and TL response compared to the bones away from the reddish patches that we here interpret as fireplaces. Bones deposited at the same time would have essentially the same response, since there was nothing to re-set on those burnt at the time of deposition. In fact, both the bones beside the fireplaces and those off in the distance give the same ESR/TL response.

But would early people with only medium sized brains be able to start their own fires? Our phytolith studies have permitted the partial identification of the woods used in the ancient fireplaces, when the fuel was from a wood species that produces distinctive phytoliths. Much of this fuel was palm wood, easily identified by the spherical palm phytoliths with numerous spiky projections. Epie Pius, an anthropologist from the Bukasi people of Cameroon, who is familiar with the various firewoods of Africa, asserts that palm wood fuel is chosen nowadays where ease of ignition and height of flame may be important. Even when dry it produces a lot of smoke. Palm wood burns rapidly, meaning that a maintained fire of palm wood requires recurrent tending. To have the on-going fires indicated by Bellomo's archaeomagnetic studies means that someone has to deliberately cultivate the hearth. The choice of palm wood seems to be partly determined (along with its availability) by its ease of ignition, despite the troublesome need to be tended. Burning a lot of palm means that someone had to look after the fires and tend them.

Chert flakes are not the only artifacts found at the ancient fireplaces. The serrated cutter shown in *Figure 11.2* was found in Reddish Spot 1 (which we now call fireplace F1), and was contemporary with it as shown by its reduced TL, from having been heated at the same time as the fireplace, as compared with other small flake found some distance from the reddish spots. (*Figure 11.3*) Near this fireplace was the mandible of a robust *Australopithecine boisei*, number KNM-ER 3230. The stone cutter may have been used to cut some meat, or even the tongue, from this mandible. The mandible had been broken at the chin, and had been placed out of anatomical order, with the right jaw on the left side, although the excavators noted the otherwise undisturbed, pristine condition of the site and everyone has commented on the sharpness of the stone tools left behind. Perhaps this dismemberment took place beside the fireplace to facilitate cooking, a scenario depicted in the painting (*Figure 11.4*). LAWRENCE KEELEY (1997) detected some cutting of soft fleshy tissues in the microscopic use wear on the tools from this site complex and cut marks were found on three bones, one of them a bovid front leg bone.

234   Ralph Rowlett

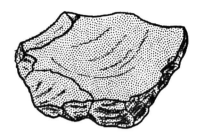 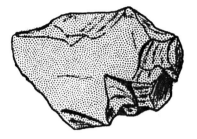

Figure 11.2. Serrated cutter found in Reddish Spot 1.

*Figure 11.3. Karari denticulated cutter found in F1 and
debitage flake found away from fireplace.*

Our current research on these ancient fireplaces has been expanded to look for evidence, if any survives, of these tubers and other plant foods in the fireplaces. We are currently studying additional basin-shaped reddish lenses, subsequently excavated in the Okote Tuff at other Koobi Fora localities by Harry Merrick (*pers. comm.* 1993), to determine which of these may be fireplaces and which may be burnt trees and/or other phenomena. The tests with Merrick's reddish patches are not yet completed. Further experiments center around the work required to keep a largely palm wood fire going as well as ignition experiments to determine how easily, or difficulty, various woods can be set ablaze. We have included woods such as those found in the *Homo erectus* bearing layers of Zhoukoudian in this study.

Circumstantial evidence also suggests that *Homo erectus/H. ergaster* was cooking food prior to eating it. As noted by RICHARD WRANGHAM and his colleagues at Harvard, likewise, the molar size is reduced and the enamel is thinner, the brain becomes bigger, the gut size (as deduced from the skeletal torso) is reduced, and women gain considerably in body size.

This same research team, noting particularly the increased mass for women, has postulated (WRANGHAM *et al.* 1981) that most of the cooking was directed toward African tubers, which would have been gathered by the women. These tubers require more cooking to be chewable and edible than does meat, so the presence of fire would have greatly expanded the food availability for these

Figure 11.4. Reconstructed scene of Homo erectus/H. ergaster cooking vegetal foodstuff
at a fireplace. Painting by Jane Bick Mudd, Department of Art,
William Woods University, Fulton.

ancient people. This enhanced contribution of women to subsistence may have
been fundamental for the organization of human society.

Some burnt materials from Zhoukoudian Locality 1 were initially examined in
1975, when Dr. Meo installed an exhibition of Chinese prehistory and archaeology
in the Nelson-Rockhill Museum in Kansas City. These hammerstones and burnt
hackberry seeds (*Celtis barbouri*) are clearly heat crazed and carbonized, so
obviously they have been burnt somewhere at some time (*Figure 11.5*). Recently
Dr. Zhao Zhiujun of the Beijing Archaeological Institute made available to me

sediments from Layers 7–11. Sediments from some portions of Layer 10, with TL dates from 417–592 years ago (Wu – Poirer 1995, 72–73), produces a TL less than that of Layers 7 and 8, the first of which has never exhibited any traces of fires and which seemingly has a TL response referring to the geological age of the sediments composing the layer. Likewise, our samples from Layer 11 also had a high TL response as the array shows in *Table 11.1.*

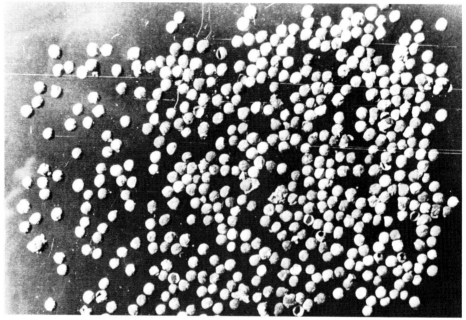

Figure 11.5. Carbonized hackberry seeds from Zhoukoudian, Locality 1.

Another offsetting advantage of burning palm is that its tall flames would help make the fires effective for frightening away predators, of which there were plenty at that time. In addition to such standard predators as lion, leopard, large cheetahs, hyenas and jackals, there were additional species of hyenas and fearsome saber-toothed felines (Brain 1981). The effusive smoking of the palm fuel would also discourage a much smaller and insidious set of predators, insects and mosquitoes.

Glynn and Barbara Isaac (1997) had already pointed out that Koobi Fora site FxJj 20E was unusual in the amount of 160 chert flakes present, in addition to the nearly 2,000 stone tools made of basaltic and other volcanic stone. Some of these flints were blackened, and some were reddish and show "pot lid" damage from having fallen into the fire. These cherts were possibly the strike-a-lights used to ignite the fires and sizeable chert flakes lay near each fireplace, but not near the burnt tree.

Table 11.1. TL Response of Selected Matrix Sediments of Zhoukoudian, Loc. 1

Layer	Response	Interpretation
7	High	No signs of fire
8	High	No signs of fire
9	High	No signs of fire
10	Low	Burning in antiquity
11	High	No signs of fire

What we were looking for in these actualistic or modern studies were phytoliths, the microscopic small silica bodies produced in the plant tissue by many species of plants. Detailed studies of 10,000-year-old charcoal in Mesolithic Danish fireplaces indicated that usually more than one kind of fuel would be used (MALMROS 1997, 170–174). Our thought was that the phytoliths of burnt trees should be much more homogenous than the phytoliths in fireplaces, which would have likely involved different kinds of sticks and woods, as well, in some instances, grass and other tinder to help start the fires. Elizabeth H. D. Rowlett processed and identified the phytoliths for us at the Palaeoethnobotany Laboratory of the University of Missouri.

The Koobi Fora fireplaces appear as inverted lens-shaped reddish patches in the Okote Tuff. Charles Peters and I used thermoluminescence (TL) to demonstrate that the reddish patches at FxJj 20 East were indeed heated more recently than the surrounding tuffs and were not the result of either fungal invasion or precipitation of iron particles. If there had been a grass fire, the geological TL response of the tuff would have been reduced to the same level as that of the reddish spots (ROWLETT – JOHANNSEN 1990; ROWLETT 1993). Of course, there was no charcoal left behind, after been leached away during the last 1.6 million years. Comparison with lightning strikes in Africa, Georgia and Kansas City showed that the 40–50 cm. "fireplaces" were not due to lightning, which seldom leaves a fulgurite with a diameter exceeding a centimeter. We burnt modern silver maple stumps in Missouri and then excavated the remains to compare the resulting burnt areas with the basin shaped reddish patches. A burnt tree tends to produce a jagged bottom outline, where the tops of some roots begin to burn, unlike the basin shaped burnt areas of Koobi Fora.

While these and other studies made it clear that the reddish spots were due to burning, we were still worried about the more difficult problem of distinguishing fireplaces from burnt trees. Glynn Isaac and HELEN KROLL (1997) had already established that the site at Koobi Fora had been open woodland, with the stone tools usually left in the shade of a tree, now long gone. In modern savanna-like open woodland at the edge of the prairie in northern Missouri, we induced some students to make some different kinds of fireplaces and to burn some trees, without telling them about the hypotheses being tested. We had the students make fireplaces with grass tinder and some without. Some of these fireplaces were made by burning several kinds of wood, and some with only one species of wood. We also had the students burn a few trees (actually stumps, because we did not want to unnecessarily kill a tree), which, would of course leave behind only one kind of wood – the wood of the tree itself. Fortunately, in this same open woodland we were testing had already been built a simple fireplace to roast hot dogs, as well some stumps had already been burnt just to get rid of them. A preliminary outcome of this project was reported in 1990 (ROWLETT 1990), and today I present still more additional information about the research accomplished since the latter publication went to press.

Our studies confirm the presence of fire and burning at Zhoukoudian. At this moment it is difficult to discern from the methods of over two generations ago if these fires at Zhoukoudian were controlled or not. However, given the compelling evidence that controlled fires were made on palm wood a million years earlier than Zhoukoudian implies that it is not unthinkable to postulate that people in the Daku glacial times at Zhoukoudian did make use fire. If the redbud and other woods burnt here are more difficult to ignite than palm, I still do not see that as insurmountable problem. By Layer 10, *Homo erectus* had had a million years practice with fire making, so it seems reasonable to suggest that pyrotechnology, like lithic technology, saw some advancement from this long experience. The fireplaces of Koobi Fora and Zhoukoudian are relatively simple, but not more so than those made by picnickers in the woodlands of modern Missouri or those made by modern foragers, such as people of the Gwich'in tribe, in northern Alaska.

Bibliography

BELLOMO, R. 1990
  Methods for Documenting Unequivocal Evidence of Human Controlled Fire
  at Early Pleistocene Archaeology Sites in East Africa. *PhD. dissertation*,
  University of Wisconsin.

BELLOMO, R. – KEAN, M. 1997
  Evidence of hominid-controlled fire at the FXJj 20 site complex, Karari
  Escarpment. In: Isaac, G. L. – Isaac, B. (eds.), *Koobi Fora V: Plio-Pleistocene
  Archaeology*. Oxford, Clarendon Press, 223–236.

BRAIN, C. K. 1981
  The Hunters and the Hunted. Chicago, University of Chicago Press.

ISAAC, G. L. – ISAAC, B. 1997
  Koobi Fora V: Plio-Pleistocene Archaeology. Clarendon Press: Oxford.

KEELEY, L. 1997
  Microwear Traces on a Selected Sample of Stone Artifacts from Koobi For a.
  In: Isaac, G. L. – Isaac, B. (eds.), *Koobi Fora V: Plio-Pleistocene Archaeology*.
  Oxford, Clarendon Press, 396–401.

KROLL, H. 1997
  Lithic and faunal distributions at eight archaeological excavations In: Isaac,
  G. L. – Isaac, B. (eds.), *Koobi Fora V: Plio-Pleistocene Archaeology*. Oxford,
  Clarendon Press, 459–538.

LOCKE, R. 1999
  "A Fiery Debate." *Discovering Archaeology*, Vol. 1, 86–87.

MALMROS, C. 1997
  *Charcoal from Stone Age fireplaces. Danish Storbaelt since the Ice Age: Man,
  sea, forest*. Copenhagen, Kalundborg Regional Museum, 170–174.

ROWLETT, R. M. 1990
  Burning Issues in Fire Taphonomy. In: Fernandez, S. (ed.), *Communicaciones
  de Reunion de Tafonomia y Fosilizacion*. University of Madrid, 327–336.

ROWLETT, R. M. 1993
  Ancient grass fires detected by thermoluminescence. *Archaeology in Montana*,
  29–32.

ROWLETT, R. M. – GRABER, R. – DAVIS, M. 1999
Friendly Fire. *Discovering Archaeology* 1, 82–90.

ROWLETT, R. M. – JOHANNESEN, S. 1990
Thermoluminescence response interference from the La Mesa fire, Bandelier National Monument. In: Traylor, D., Hubbell, L., Wood, N. – Fielder, B. (eds.), *The 1977 La Mesa Fire Study*. Santa Fe, New Mexico, Southwest Cultural Resources Center, 191–201.

SILLEN, A. – BRAIN, C. K. 1990
Old Flame. *Natural History*, 4, 6–10.

WRANGHAM, R. – HOLLAND JONES, J. – LADEN, G. – PILBEAM, D. – CONKLIN-BRITTAIN, N-L. 1999
The Raw and the Stolen: Cooking and the Ecology of Human Origins. *Current Anthropology*, 40: 567–594.

WU, X. – POIRIER, F. 1995
Human evolution in China. Oxford, Oxford University Press.

12

Pyrotechnology and Landscapes of Plenty
in the Northern Boreal

ROBERTA ROBIN DODS

Introduction: Wilderness and the Savage

From the late 16th Century onward émigrés to North America arrived with certain concepts of landscape that were to have long-term and telling consequences. Their views of nature constructed physical and metaphorical North American landscapes (VON MALTZAHN 1994) familiar to us today in the primeval forest myth I have discussed elsewhere (DODS 2002). As early as the 1630s the forests of eastern North America were described as a "desart wildernesse", this in the sense of deserted places (BOWDEN 1992, 5–6). By defining North American landscapes as deserted, concomitant with the inverted versions of the savage as noble/ignoble, indigenous peoples became both figuratively and literally the disappeared. Additionally, 'wilderness', by definition devoid of people or at least 'real' people – the civilized, needed to be "replaced with a structure governed by the logos" (VON MALTZAHN 1994, 126). The application of the logos of the colonial power was a weapon of annihilation of the indigenous oral tradition world leaving a blank 'page' on which the economic objectives of Euro-Americans[1] could be written. Thus a translation had commenced of North American oral kin-based societies situated in traditional domesticated landscapes into English of the Early Modern Period on its way to a capitalist vocabulary. Such a translation never included the aboriginal peoples as managers of a rationally constructed and managed culturally defined physical landscape.

Early European reports (DE VRIES 1911 [1655], 154)[2] on the burning of eastern North American forests were dismissed as "isolated accounts". It is true that these

---

[1] Euro-American is used to indicate North Americans of European ancestry (what are now called Americans and Canadians).

[2] See also JAMESON (1909) and PARR (1969); Rosier's 1605 journal entry on his three hour hike through the tall treed forests of the St. George River area of Maine (QUINN 1973, 390; THOMAS 1774).

early descriptions are not value free in their writing or even in our reading. The content, referent, and context are more or less lacking from the original records and we are pushed by whatever agenda is current in our discipline to find certain meanings in the record of the past. However, if such accounts are examined in the context of archaeological evidence tied to modern data on forest ecology a startling and challengingly different picture emerges. This picture moves us beyond the idealization of the *forest primeval* as a wilderness, an empty land, an unmanaged landscape devoid of significant and telling cultural constructs.

Additionally, population collapse from introduced diseases masked traditional forest constructs since the removal of the traditional managers and their management techniques, meant that the forests increasingly became old growth. A particular example of this can be found with the smallpox epidemics in the northern boreal of the 18th century (UMFREVILLE 1790, 92–93). Later historic description such as PARKMAN's of 19th Century New Hampshire, portrayed unmanaged forest as being of "decrepit age" full of the "impotence of rottenness... devouring its own dead" (1865–1892 [1983]; 1, 1322). Thus the problem persists in the 'reading' of these forests the vocabulary of meaning remains foreign to the cultural outsider as observer who, in the past, used "immediate perceptions...[and] relied upon implicit and at first unchallenged beliefs..." (PAGDEN 1982, 11) about the nature of such landscapes.

Maintaining the image of the North American forests as primeval wildernesses meant that it became almost impossible to recognize the boreal forests of North America as being the result of centuries on centuries of indigenous management practices that shaped landscapes to both economic and spiritual needs. In this chapter I will examine the place of pyrotechnology and keystone species in the development of what I term *landscapes of plenty*, essentially domesticated spaces of the past, in boreal North America.

To transcend the Euro-American idealized view of the past there are a number of tenets that must be accepted.

1) The criteria for the development of sylvan domesticated landscapes in North America differed from those for sylvan domesticated landscapes in England specifically and Europe in general.

2) Regardless the immediate devastation caused by fire, the long-term effect of cyclical fire regimes on the habitat can be crucial to the overall carrying capacity of specific environments.

3) Generally, boreal forests of North America are fire generated-regenerated (DELCOURT – DELCOURT 1987; 1981) and "In the north most natural forests are either maturing following the last fire or being instantly recycled by the next" (HEINSELMAN 1981, 386. Emphasis added).

4) Knowledge of fire regimes can be used to generate through pyrotechnology forests with high carrying capacity of specific resources.

5) The forests of the 'New World' were constructed landscapes with their own histories as much as the forests of the 'Old World'.

6) These forests were, and remain, even in altered circumstances of today, a text to be read for social content as much as the coppiced and enclosed landscapes of an England long gone.

The 'reading' of these landscapes can be done from varied perspectives, some more powerful than others. The acceptance of the observations from the initial contact period can be tied to the examination of modern analogue environments. Insights from archaeological faunal data supplement such observation (DODS 1998). It is through the synthesis that the various forms of inquiry come to reveal the organized spaces that constitute landscapes constructed through human agency.

Hudson Bay Company Moose Factory, James Bay Log Books 1730–1820

The cogent feature of the Hudson Bay Company logbooks is that they read like ship's logs where weather is a constant theme. Written as an account of daily activities in the forts, they were meant to inform the Governors of the Company in England of the developing Factories (forts) and their commerce. Lightening storms are noted but they make little mention of fire either wild or domestic. Now, in all fairness, it must be said that the documents are relatively devoid of cultural observations in their initial years. In the period of a Factor called Eusebius Bacchus Kitchin, around the 1770s, the moose records improved but not to any ethnographic standard.

However, an analysis of the information contained in these logbooks leaves a number of questions. Why is there no comment on the nature of the landscape needed to sustain their major trade objective, namely the pelts of the American Beaver? Certainly they wanted furs of quality generated from winter hunts but beyond this there is almost a compete lack of comment on what we would term the ecological conditions to sustain a high harvest. Such is the case that Kitchen's

(1774–1775) comment of December 20, 1774 that a "...man who came in on Sunday informed me that he had found a beaver house contiguous to the fort and as I never as yet seen one taken, notwithstanding my long residence in the country, I purpose going tomorrow and eight men with me to assist the Indians, to make dispatch,..." stands apart in general tone. The goose hunts that provisioned the fort were covered in some detail and the numbers from the spring and fall hunts illustrated the long-term cycles in goose population densities.

If the forest was unmanaged and basically "ancient and wilderness", in other words in late stage growth, then why did it not burn? There are two comments on clearing brush away from the fort because of the possibility of fire, one comment on smoke to the north that is seen but that one day and basically that is it. If fires occurred, why were they not a matter of comment? Why did lightening strikes not explode into the large intense wildfires so dreaded in old growth forests? Certainly lightening strikes were noted but these were not apparent producers of extensive fires. This seems relatively strange from my experience in the field season of 1995 in northern Saskatchewan where 92 forest fires were burning in my study area.

Another feature of the fort records is the constant and ongoing comment about the production of lumber for fort constructions projects as well as the maintenance of the various boats used for regional communications, ferrying to and from the ocean-going ships from Britain, and for occasional forays of exploration along the coastline. Here we see described in some detail the economic use of the forest in the European model – forests as producers of wood. Yet in none of the observations on the cutting of trees and the removal of forest products to the fort is there mention of the difficulties of transport through impassable forests or of forests marked by fire.

These daily entries can be set against a document embedded between two Moose Factory Log pages for the month of July 1774 (*Appendix 1*). It is within the context of these mundane fort logs that the Thomas journal becomes all the more remarkable. These 'remarks' of JOHN THOMAS are a mere 16 pages on microfilm (B.135/a/55) consisting of 32 handwritten pages of the daily log entries, on Thomas's trip in the summer of 1774 from Moose Factory on James Bay to 'Abbitibby' in the interior of what is now eastern Northern Ontario. This document contains interesting cultural and environmental observations including descriptions of burn areas. On 16 of the 36 travelling days he comments at least once and not infrequently twice on fire areas. Some of the fires appear to have been very large. For example on July 13th, when he estimated they had travelled

15 miles, Thomas states "most of this day all burnt woods". This after the 12th, a ten mile day, when he commented "burnt woods" early in the entry and then later noted "when we put up, still burnt woods". Are the burnt areas of the 12th / 13th may the remains of the same extensive fire or a series of patches running into each other or what have been described as corridor burns? We do not know the answer and, further, we have no way of finding the answer from the current investigative techniques available to us.

Some of the Thomas entries suggest patch burning such as LEWIS and FERGUSON (1988) describe for northern Alberta in modern times. For example Thomas states "burnt in many places" (July 25th), "some places burnt" (August 12th). Another entry suggests a ground burn with "1/2 burnt woods" (August 2nd) and certainly my own fieldwork in Saskatchewan indicates that these 1/2 burnt areas are optimal grazing locations if the fires occur early enough in the season. Interestingly, this area south of Moose Factory is characterized as "fire territory" (RICH 1961). Furthermore, Thomas makes note of beaver lodges and dams. On July 6th it is "beaver dens" and the next day it is "many beaver tracks & houses" and so it goes. On July 28th he comments on three beaver dams made that summer and the nearby "several old houses". Here is a record of the relationship between fire and beaver that is central to the prehistoric landscape management objectives.

Fire Ecology Observations from Modern Environments

The fieldwork in modern boreal environments was conducted in central-northern Saskatchewan in the summer of 1995. Selection of six sample sites was made from the infrared images provided by the Boreal Ecosystem-Atmosphere Study (BOREAS) for their Southern Study Area (SSA)[3]. Fives sites were used for data collection, the sixth was unavailable because of recent flooding. The western sites were in Prince Albert National Park (PANP) and the eastern sites in Narrow Hills Provincial Park (NHPP). Selection criteria were based on the 'Image Value' (1–11) of the classes of ground cover seen in the infrared images (*Table 12.1*). As a matter of course 1995 turned out to be one of the worst fire seasons in recent years in Saskatchewan. Recently extinguished or still burning fires provided excellent

---

[3]   Access to this material was facilitated by Prof. Peter Muller, and his research assistant Tim Wilkinson of Photogrametry and Surveying, University College London, England and by Dr. Forrest Hall of National Aeronautical and Space Agency (NASA), United States of America.

locations for collection of data on very early stages of regeneration. Further, the areas chosen for fieldwork had excellent documentation on the forest fires of past years.

Table 12.1. Classes of ground cover for the SSA 129 km by 86 km north of Prince Albert Saskatchewan (BOREAS 1995).

CLASS	AREA m^2
Conifer (Wet)	5,351,030
Conifer (Dry)	87,031
Mixed (Coniferous and Deciduous)	1,938,822
Deciduous	1,085,214
Disturbed	130,461
Fen	417,206
Water	1,086,167
Regeneration (Medium)	315,947
Regeneration (Younger)	149,492
Regeneration (Older)	716,549
Visible Burn	41,718

Generally the southern section of both study areas are ecotoning from the great Canadian prairies increasingly to the boreal forests as one moves north. PANP is an area of gentle relief while NHPP is named for the long narrow hills that provide upland areas between lakes, rivers and the expanses of muskeg. The rivers and lakes are affected by the surface geology created from late Pleistocene post-glacial fluvial-lacustrine action. Beaver are ubiquitous. Interspersed are modern beaver environments and grasslands, some the direct result of ancient beaver ponds having developed into water meadows then drier environments.

PANP last major fire was in the 1940s, although there have been small local fires that have been promptly extinguished as was the policy of forest fire suppression by the Canadian Forestry Service. My cynical analysis was that the objective of this policy was to keep the forest in a structure that "fits" the landscape image Euro-Canadians have of the wilderness, the forest primeval, the virgin forest. A self-renewing fire-cleaned forest would cause the "problem" of too many elk; "we'd be over-run by elk" was the exact comment. In traditional systems would not "too many" animals be the goal?

In PANP the majority of the youngest trees are a mere fifty years old, in second stage regrowth. Older trees can be over one hundred years old but mostly they represent the regrowth from the major cuttings during the time of the western expansion of the Canadian railway system in the 1890s. There are a few rare areas where the last fire was in the 18th Century and the trees escaped the lumbering of the last century. All logging ceased in the early 1950s. These conditions do not pertain for NHPP where parts are under lumber – pulp license and of course burn tracts picked over by the companies that hold concessions in the area. NHPP was in various stages of regrowth from a series of four major and numerous minor fires between 1977 and 1995. There is some fire edge overlap between fire areas where first stage regrowth from a pervious fire is lost to a later fire. The significance here is that a recent fire zone provides very little fuel for a subsequent fire and thus may act as a fire-break zone if the next fire is not driven by high winds during very dry, hot weather.

Modern industrial reforestation does not reproduce equivalent environments as found in a natural successional post-fire regime (BØRSET 1976). In this 'modern' system undergrowth becomes 'weeds' while many animals are defined as 'pests' and dealt with as such (BALTENSWEILER – FISCHLIN 1987, 401–415). The principle of community is violated by such reforestation, which encourages a "levelling out" of the characteristic vertical mosaic of the boreal forest. Diversity of undergrowth plant species that associate with certain tree species complexes are not sustained (HEINSELMAN 1981). This undergrowth affords browse for the herbivores, large and small. Even under 'natural' conditions there may be subsequent drastic environmental changes brought with regrowth. Yet such natural shifts offer more diversity than we find with today's industry-based programs of reforestation, better called plantings, for a forest is a sum of many parts not a monocultural statement. Post-fire forest regrowth fidelity will depend on a number of factors including:

- the type of fire (crown, surface, or ground) and intensity (based on fuels and wind);
  the season of burn;
- the predominant species composition of the initial forest;
- percentage of shrubs/herbs/grasses adapted for vegetative reproduction;
- rainfall amount and timing;
- the survival of seed beds; and
- percentage of destruction of incipient regrowth by mammalian browsers (HEINSELMAN 1981, 377, 387, 390–391).

The decision to burn a tract of land would entail making choices about the type of burn and the outcome desired. Summer-early, autumn burns would be avoided since this is the most dangerous 'fire season' of the boreal forest. For old, deep forests of great extent the danger of an out-of-control fire would have been too great in the heat of the warm months. Such areas could be set afire late in the autumn and left to burn as a group left an area for another camp location. The coming cold weather and snow would limit such fires to restricted localities. However, this was probably not needed that frequently in the managed landscape of the prehistoric boreal world. The main objective in the management system being considered here would be a low intensity, easily controlled fire that would clear out the dead and encourage new growth. Such environments would maintain early successional characteristics with the patterns of use of beaver, and to a lesser extent other browsers and grazers. Two seasons would be suitable for this, spring and later autumn. Further the burns in each of these seasons would have slightly different outcomes:

Firstly, the spring burn has advantages as seen in PANP research in 1995 (DODS 1998).

They are low intensity burns that clean out old undergrowth while thinning, but not destroying, the leaf cover. The result is the generation of new lush undergrowth in which browsing and grazing can occur.

The autumn burn entails more danger than the spring burn when ground cover is still lush with moisture. Normally autumn is a dryer season and follows directly on the usual fire season of the boreal area. At this time fires are more intense since the living biomass is dryer and dying back it contributes to the fuel load. Fires are more likely to be crown fires that cause serious damage to the trees. There will be very little regrowth since winter is so near and food resources for certain animals may be destroyed or seriously decreased by burning in the late autumn (e.g. browse for large herbivores such as deer and moose). It can be used to clear a large section for the development of extended open areas that will be productive in the future. However, it is not the strategy for immediate returns.

One can gain a sense of the emerging post-fire world by looking at the preferred habitats of a few of the important economic species from the prehistoric record. Because of fauna habitat preferences, selected species would be more abundant in incipient regrowth areas because of altered parameters for optimal populations. Here, then, is an important issue if understanding of the past is to be developed by application of observations from today. The species that feed on early successional regrowth can retard the eventual development of the climax forest situation. Therefore the system is somewhat self-perpetuating once in operation. Thus beaver, moose, hare, deer and other animals that are early successional herbivores can, by their feeding patterns, recycle the early stages of boreal climax development and influence the eventual species composition of the emerging post-fire conifer forest. Such retardation of forest development would lengthen the cycles of the natural fire regimes and incidentally, or perhaps purposely, suit the prey acquisition needs of humans. In this environment Beaver are keystone species (NAIMAN *et al.* 1986).[4]

The environmental requirements of beaver are complicated and critical to the development of the optimal economic landscapes of prehistory. Beaver numbers are closely tied to water levels of sufficient depth as to discourage the growth of bacteria that causes the disease *tularemia*. Slow, meandering streams and creeks bordered by secondary regrowth aspen and birch, as well as lakes that are fed by or feed into streams, are the preferred habitat (BICE 1983, 101–102). Observations suggest that beaver ponds can cover as much as 13% of the land area in this part of the world. In addition, there are the heavily browsed zones that surround the ponds (PASTOR – MLADENOFF 1992, 232). This is not random behaviour. The selection by Beaver of early-successional hardwoods such as birch and aspen is to be expected "...because these species contain lower concentrations of carbon-based, secondary compounds than do conifers, or produce these compounds only during juvenile phases" (PASTOR – MLADENOFF 1992). These carbon-based compounds act as defences against herbivore predation. How these anti-predation compou nds contribute to the C:N ratio in the forest could be the topic of another paper but sufficient to this presentation is the fact that selective browsing "...shifts

---

[4]   The concept of 'keystone-species' was first introduced in 1966 by R.T. Paine's study of intertidal systems' food web complexity and species diversity. However it was not until 1969 that the Paine actually applied the term of 'keystone' in a short article on trophic complexity that appeared in *American Naturalist*. There are "...two hallmarks of keystone species. First, their presence is crucial in maintaining the <u>organization</u> and <u>diversity</u> of their ecological communities. Second, it is implicit that these species are exceptional, relative to the rest of the community, in their importance..." (MILLS *et al.* 1993, 219) (emphasis added). The authors also state an interesting fact with respect to the management of keystone species noting that they "...could be used to support populations of other species..." (MILLS *et al.* 1993, 219). A classification of keystone species was eventually developed. Eventually five categories of keystone species were suggested: keystone predator; keystone prey; keystone mutualists; keystone hosts; and keystone modifiers (MILLS *et al.* 1993, 220). It is to the final category of keystone modifier that *Castor canadensis* (Beaver) belong.

competitive balance further toward conifers..." (PASTOR – MLADENOFF 1992, 233). Thus the landscape may be shaped by herbivore selection of prey plant species and the developing spatial patterns in nutrient cycles (PASTOR – MLADENOFF 1992). Thus Beaver are central to the analysis of prehistoric economic patterns. They are effective and significant shapers of both open and bounded aquatic systems. This in turn impacts upon the land-based portion of the environment. As such they generate diverse habitats that support other species that are also part of the prehistoric food economy. Their early successional requirements can be initiated by the use of fire.

It is the beaver who turns open aquatic systems into extensive bounded systems such as marshes and swamps. These in turn provide more wetland habitats for the waterfowl and generate more edge areas for the edge feeders such as moose and deer. The removal of beaver and the elimination of their landscape modification effects cause these bounded systems to dry out. Typically these desiccated wetlands initially become extensive open areas that provide feeding meadows and extensive edge areas for land-based mammals such as hare. Deer also use such areas for grazing portions of their diet. Eventually some of these areas will evolve to closed forest systems that have their own ageing processes through to the old growth stage. Both water and land areas, as well as their ecotones, are important for *cervidae* such as deer and moose since these species require very diverse habitats consisting of meadows and parklands, swamps and marshes and ecotone areas abutting first and second stage forests. Such complex habitats can be initiated and maintained by beaver damming activity and, to a point, the various stages of landscape ageing of open aquatic systems.

The meadow and edge areas that are the spin-off of the habitat alteration of beaver (or are the result of re-established environments after a fire) offer habitat to the hare as well. Hare are modifiers of environments although generally not to the extent of beaver. This is an animal noted for wide population fluctuations. The swings between peaks and troughs of population amplitude, which are of the order of 3400:1 in northern Canada, are spaced six to thirteen years apart (mode 9–10 years). The peaks do not appear simultaneously and neither do the troughs in all areas of an ecosystem or over wide expanses of territory. Further, the variations in population are not of equal measure across wide geographical areas (BANFIELD 1974, 82). This means that humans have the opportunity to use resources in adjacent areas when that area maintains longer or rebounds sooner. In this case kinship relationships and marriage alliances (e.g. cross-cousin marriages that link over patrilineages (1929), brother- sister/sister-brother marriages, and hunting partnerships would allow access to abundant resources found in the territory of others). This same principle of territory sharing within and between kinship groups would also function when pyrotechnology was used to reset a specific environment to first stage regrowth.

Beaver were consistently number one in the mammalian faunal remains at the archaeological sites of Martin Bird, Wabinosh and Long Sault analysed from Northwestern Ontario (DODS 1998). This is the case even in the face of the low pH readings and the expectation that smaller bones and bones from smaller animals would be more vulnerable to loss in acidic soils than larger bones and bones from larger animals.

However, being the smaller animal does not mean that it is less productive per unit of space than the larger animal, as will become significantly evident below. Beaver have certain reproductive traits that group them with the muskrat, hare, and even deer in their ability to adjust reproduction up or down in response to the conditions they experience in their own portion of the environment.[5] This is understood by the boreal hunter. It is part of traditional ecological knowledge and as such can be incorporated into management decisions on standing crops, hunting practices and landscape development.

The number of beaver for North America at the time of contact has been estimated at 60×10^6 to 400×10^6 (NAIMAN *et al.* 1986, 1254). This is calculated as being the reduced figure of 6×10^6 to 12×10^6 for the 20th Century (NAIMAN *et al.* 1986, 1254). These significantly lower numbers are the direct result, in my opinion, of over hunting, human infringement on territory, and altered forest management practices where fire suppression became a key strategy (*Table 12.2*). Using these figures, the area of North America[6] in hectares (24×10^8), and the average weight of a beaver as being 20 kg[7] we can estimate standing crop biomass for these periods of time:

---

[5] Beaver, although monoestrous, can adjust the litter size through embryonic absorption; this can account for as much as 27% of prenatal mortality (BANFIELD 1974, 161). It is the result of stress on the female at a specific time of gestation. This stress can be the result of poor physical condition of the female or the scarcity of food since impregnation occurs between mid-January and mid-March, the period when Beaver primarily rely on their caches of stored food. Size of litter ranges from one to eight (average 3.9); larger litters are usually found with older mothers (BANFIELD 1974, 161; BICE 1983, 102).

[6] 24,000,000 sq km times 100 (number of ha per sq km).

[7] BANFIELD notes that weight is highly variable and depends on "age, sex, and season" and adult weight ranges from 15 to 35 kg or the average of 20 kg (1974,158).

Table 12.2. Estimated standing crop biomass for Beaver

	60×10^6	400×10^6	6×10^6	12×10^6
Animals per ha	0.025	0.166	0.0025	0.006
kg per ha	0.5	3.32	0.05	0.120

Beaver at 60×10^6 produce 0.5 kg per ha and the astonishingly productive figure of 3.32 kg per ha at 400×10^6. Although a relatively small animal, this is much more productive per unit of land than Caribou at 0.02 kg per ha (SIMKIN 1965, 742). Beaver would similarly compare favourably against deer that are at their northern limits in the boreal forest. Taking the figures supplied by PETERSON (1955, 75–77, 202–203) for moose weights and population densities. Moose would produce approximately 0.56 kg per ha. Since moose populations "... require comparatively large minimum area per animal, and the population does not normally increase beyond this density in spite of optimal habitat conditions... (or they increase) ...as temporary response to particularly favourable habitats" (PETERSON 1955, 196). Beaver compare favourably against the very large mammals of the region.

Conclusion

Pyrotechnology as a management tool expanded edge areas, encouraged plant communities that flourish under frequent fire regimes and developed habitats for animal communities that associate with early successional environments (MACCLEERY 1994, 4). The additional benefit of having areas that are early successional is that they are fire resistant. In the advent of wildfire from a lightning strike or campfire accident there is the survival advantage of the planned sanctuary from such dangers.

Beaver loomed with a significance that is telling to this analysis because:
- it was, literally and figuratively, central in the life of these people – Beaver was a key metaphor;
- it is relatively well represented in the pre-contact archaeology record (DODS 1998);
- it became central to the post-contact economy through an accident of history and was the species of survival in the newly emerging commodity market economy with the Hudson Bay Company;

- it was one of the species with a malleable reproductive strategy. This caused it to be central to the traditional subsistence economy through its position in a managed system;
- it had the most effective interaction with the second strategy used by humans – the resetting of the successional sequence through the use of fire as a management tool; and
- thus as a keystone species, Beaver was strategic to the survival of the ecological relationships in this area of North America.

Since vegetation may be shaped by herbivore selection of prey plant species and the resulting spatial patterns of nutrient cycles caused by this selection, early successional herbivores feeding patterns cause the initial stages of boreal forest climax development to recycle. Thereby herbivores influence the eventual species composition of the emerging post-fire conifer forest and the relationship of open areas to closed areas. Concomitant, based on their habitat preferences, various species of mammals would be more abundant in post-fire incipient regrowth areas because of the altered parameters for optimal populations, for example hare, *cervidae* species and moose.

Other species, such as woodland caribou, that rely on the old growth understory of lichens and mosses, would be rare or absent. Further, the specific example of beaver is important for the analysis of cultural activity in prehistoric boreal North America. This animal participates in the shaping of early stage landscape and the development of wetland habitats that cycle through to open grassland areas. It is these grassland areas that, in turn, support additional economically useful species listed in the first instance, above (hare, deer and moose). "Feedback" systems between habitat and specific animals would have suited the prey acquisition needs of humans and humans could have encouraged such systems by use of certain strategies, including fire, to create the "start-up" and maintenance of such productive areas. An additional benefit would be an increase in production of certain berry bearing plants such as blueberry (*Vaccinium* sp.) (GOTTESFELD 1994; VEIJALAINEN 1976).

Once on the road to a market economy future, it became impossible for the indigenous peoples to turn back. Elders who were the repositories of traditional knowledge were dead or dying, or they and the young were choosing to look to new technologies instead of old ones. The young were losing their guides and/or their own knowledge to traditional technologies and thereby the return route to a completely traditional culture. In this transition from the traditional subsistence economy past to the market driven commodity future, the beaver became increasingly needed as the medium of exchange. Eventually the trade included non-traditional foodstuffs and this shift in diet, albeit some of it was enforced by the consequences of the demise of the game animals in the region, has been the broken link to the traditional past that has had the widest ranging consequences to the continued health of northern peoples (FARKAS 1979). Additionally in the historic period living lightly on the land was interpreted as non-use and Euro-Americans failed to see that the forests of North America were originally domesticated landscapes organized from a regenerative food economy biomass perspective where animals not lumber were the objective and fire was a tool whereby these economic objectives could be maintained.

Acknowledgements

Modern boreal environment research was supported by the V. Gordon Childe Fund, Institute of Archaeology, University College London (UCL) and the Graduate Research Fund, UCL. Sample sites selection was made from the infrared images provided by the Boreal Ecosystem-Atmosphere Study (BOREAS) for their Southern Study Area (SSA) facilitated by Prof. Peter Muller, and Tim Wilkinson of Photogrametry and Surveying, UCL and by Dr. Forrest Hall of NASA. The background work on early modern Britain was completed with the support of the Department of Archaeology, University of Durham, England while on sabbatical from Okanagan University College, Canada.

Bibliography

BALTENSWEILER, W. – FISCHLIN, A. 1987
    On methods of analyzing ecosystems: lessons from the analysis of forest insect systems. In: Schultz, E. D. – Zwolfer, H. (eds.), *Ecological Studies, Vol. 61, Analysis and Synthesis Potentials and Limitations of Ecosystem Analysis*, Berlin, Springer-Verlag, 401–415.

BANFIELD, A. W. F. 1974
    The Mammals of Canada. Toronto, University of Toronto Press.

BICE, R. 1983
    Fur: The Trade That Put Upper Canada On The Map. Toronto, Ministry of Natural Resources, Ontario.

BOREAS 1995

BOREAS Supersite. Section 10–1. At <http://boreas.gsfc.nasa.gov./>

BØRSET, O. 1976

Introduction of exotic trees and use of monocultures in boreal areas. In: Tamm, C. O. (ed.), *Man and The Boreal Forest.* Stockholm, Ecological Bulletin (Stockholm) 21. Proceedings of a regional meeting within Project 2 of MAB (UNESCO's Man and the Biosphere Programme), 103–106.

BOWDEN, M. J. 1992

The Invention of American Tradition. *Journal of Historical Geography* 18(1), 3–26.

DELCOURT, P. A. – DELCOURT. H. R. 1981

Vegetation maps for eastern North America: 40,000 Yr. BP to the present. In: Romans, R. C. (ed.)*, Geobotany II*, New York, Plenum Press.

DELCOURT, P. A. – DELCOURT, H. R. 1987

Long-term forest dynamics of the temperate zone. *Ecological Studies: Analysis and Synthesis*, Vol. 63. New York: Springer-Verlag.

DE VRIES 1911

Korte Historiael ende Journaels Aenteckeninge Van Verscheyden Voyahiens in de Vier Deelen des Wereldts-ronde, als Europa, Africa, Asia, ende Amerika. Gedaen David Pietersz. De Vries Uitgegeven door Dr. H.T.Colenbrander 'S-Gravenhage: Martinus Nijhoff.

DODS, R. R. 1998

Prehistoric Exploitation of Wetland Habitats in North American Boreal Forests. PhD. Dissertation. London, University College London, University of London.

DODS, R. R. 2002

The Death of Smokey Bear: the ecodisaster myth and forest management practices in prehistoric North America. *World Archaeology* 33(3) (Feb. 2002), 475–487.

FARKAS, C. S. 1979

Survey of Northern Canadian Indian Dietary Patterns and Food Intake. Man Environment Studies. Waterloo, University of Waterloo.

GOTTESFELD, L. M. 1994
Aboriginal Burning for Vegetation Management in Northwestern British Columbia. *Human Ecology,* 22(2), 171–188.

HEINSELMAN, M. L. 1981
Fire and succession in the conifer forests of North America. In: West, D. C., Shugart, H. H. – Botkin, D. B. (eds.), *Forest Succession: concepts and application.* Berlin, Springer, 374–405.

HUDSON BAY COMPANY MOOSE FACTORY LOG BOOKS, 1720–1830
Winnipeg, University of Manitoba, Hudson Bay Company Archives.

JAMESON, J. F. 1909 (ed.)
Narratives of New Netherland 1609–1664. New York, Scribner's Sons.

LEWIS, H. T. – FERGUSON, T. A. 1988
Yards, Corridors, and Mosaics: How to Burn a Boreal Forest. *Human Ecology,* 16(1), 57–77.

MACCLEERY, D. 1994
Understanding The Role The Human Dimension Has Played In Shaping America's Forest And Grassland Landscapes: Is There a Landscape Archaeologist in the House? At http://forests.org.gopher/biodiversity/landscape/txt

MILLS, L. S. – SOULÉ, M. E. – DOAK, D. F. 1993
The Keystone-Species Concept in Ecology and Conservation. *Bioscience* 43(4), 219–224.

NAIMAN, R. J. – MELILLO, J. M. - HOBBIE, J. E. 1986
Ecosystem Alteration of Boreal Forest Streams by Beaver (*Castor canadensis*). *Ecology* 67(5): 1254–1269.

PAGDEN, A. 1982
The Fall of Natural Man: The American Indian and the origins of comparative ethnology. Cambridge, Cambridge University Press.

PARKMAN, F. 1983 [Reprint 1865–1892]
France and England in North America. Two Volumes. New York, Literary Classics of the United States, Inc.

PARR, C. M. 1969
The Voyages of David De Vries Navigator and Adventurer. New York, Thomas Y. Crowell Company.

PASTOR, J. – MLADENOFF, D. J. 1992
Southern boreal-northern hardwood forest border. In: Shugart, H. H., Leemans, R. – Bonan, G. B. A. (eds.), *Systems Analysis of the Global Boreal Forests*. Cambridge, Cambridge University Press, 216–240.

PETERSON, R. L. 1955
North American Moose. Toronto, University of Toronto Press.

QUINN, D. B. 1973
England and The Discovery of America, 1481–1620. London, George Allen & Unwin Ltd.

RICH, E. E. 1961
The Hudson Bay Company 1670–1870 : Volume I:1670–1763. New York, The Macmillian Company.

SIMKIN, D. W. 1965
Reproduction and productivity of moose in Northwestern Ontario. *Journal of Wildlife Management* 29, 740–750.

STRONG, W. D. 1929
Cross-cousin marriage and the culture of the Northeastern Algonkian. *American Anthroplogist* 31, 277–288.

THOMAS, J. 1774
Mr. John Thomas's Remarks on His Journey to the Abbitibby Settlement and back again. Hudson Bay Company Document B135a 55. Winnipeg, Hudson Bay Company Archives.

UMFREVILLE, E. 1790
The Present State of Hudson's Bay. Printed for Charles Stalker, No. 4, Stationers Court, Ludgate-Street.

VEIJALAINEN, H. 1976
Effect of forestry on the yields of wild berries and edible fungi. In: Tamm, C. O. (ed.), *Man and The Boreal Forest*. Stockholm: Ecological Bulletin (Stockholm) 21. Proceedings of a regional meeting within Project 2 of MAB (UNESCO's Man and the Biosphere Programme), 63–65.

VON MALTZAHN, K. E. 1994
Nature as Landscape: dwelling and understanding. Montreal and Kingston, McGill-Queen's University Press.

APPENDIX 1

CHART OF THE THOMAS (1774) OBSERVATIONS

1774	Flora	Fire	fauna	Culture, etc.	Miles
04/07					6
05/07	Fine high woods		black bear tracks		10
06/07	Fine high pine woods		beaver dens otter tracks shot black bear	found old canoes	20
07/07	Pine woods	1/16 mile	many beaver tracks & houses		10
08/07	High pine woods	100 yards	saw deer, beaver, geese tracks		10
09/07	Pine woods Later in day High pine woods		many beaver tracks few deer tracks		10
10/07	"english firr" later in day high woods	burnt woods	"vast many beaver tracks" after fire comment		10
11/07	"green woods" "english firr"		'saw many beaver tracks"	Indian graves	5
12/07	Hour after burn: small lake with "rushes & willows"	burnt woods "when we put up, still burnt woods"			10
13/07		"most of this day all burnt woods" "heavy thunder & lightening"	"saw a vaste many beaver tracks & houses"		about 15 "half way"
14/07	"high woods around lake"	the north side burnt many years ago	"saw many beaver tracks and houses"	on lake Mis,sa,ka,mees ("cou'd not see the land to the ne end of it, many islands in it") met with Quo,te,tns and Me,tock,o,ho "here fishing with their familys"	15
16/07				returned to Mis,sa,ka,mees looking for new guide as indians from fort do not know beyond this lake. met with pufso & his family (9 in number) – one of his young fellows will be guide	
22/07				blowing	5
23/07				blowing	
24/07					15

1774	*Flora*	Fire	*fauna*	Culture, etc.	Miles
25/07		"burnt in many places"			10
26/07	*High woods*				8
27/07	*High woods*			met with *Coo,che,rau*, and family 5 in number	20
28/07	Carried canoes through the woods a number of times		"mile carriage & came into small creek (here was three dams acrosss the creek done by beaver this summer and several old houses)"	*Coo,che,rau*, drew 'a plan of the roads we have to go' the road we have come and are going the Indians call the *Mis,sa,ka, mee* path it formerly has been much used but seemingly this year has not been trod by a single individual as the grass is grown in the paths and theres no tracks of any body or tenting places, there is still standing beacons with birch rinds on the top pointing the road to the settlement"	30
29/07	Pine woods After burn comment Poplar, asp and willows	¾ way burnt			50
30/07		all burnt woods			35
31/07		burnt woods ½ way		saw three mountains about eight miles to the eastward	35
01/08				saw a large mountain to the se and another nw both of vast height	30
02/08	High woods ½ burnt	½ burnt woods		saw several winter tenting places & found an old canoe	30
03/08		some other burnt woods			
04/08			*heard gun to sw* saw canoe laying on the shore and birch rind hung up to a tree it proved to be *Acooms* son he calls himself *Acoom* (father dead)		paddled back across the lake

1774	*Flora*	Fire	*fauna*	Culture, etc.	Miles
05/08				met Indian man M*us,ke,mote*, a boy, 4 women & six children note comment on this day re Englishman father and Indian mother for a specific trader	padling back the same way we came 'till night
06/08				came to settlement and there follows long discription of area and conditions	
08/08			comment on use of dogs as food		
09/08	Gathered hazel nuts which were not ripe			start of return trip	
10/08		*burnt woods*		met with *Co,chee* and family six in number two women & 3 children	
11/08	High woods Then 4 miles of willows	*some burnt*	saw many beaver tracks		
12/08	High banks, pine and poplar woods	some places burnt			
13/08			*fresh deer meat*	see long description of encounter	
14/08	High banks with green woods Hazel nut (not ripe) gathered			as above	
15/08	Low green woods				
16/08				came to Moose River we have been 100 hours coming down from the settlement, 82 of which were with the stream over small falls of water and by allowing for the obstructions of falls . I suppose we have come 3 ½ miles per hour at least which is 350 miles and that I take to be the distance of the Abbitibby	

Participants and their Affiliation

Kevin Andrews (kandrews@bmth.ac.uk),
Department of Archaeology, University of Bournemouth, England

Françoise Audouze (audouze@mae.u-paris10.fr),
Centre de Recherches Archaeologiques, Nanterre, France

Roberta Robin Dods (rrdods@ouc.bc.ca),
Department of Anthropology Okanagan University College, British Columbia,
Canada

Marie-Chantal Frère-Sautot (mc.frere.sautot@saprr.fr),
Société des Autoroutes Paris-Rhin-Rhône, Archéologie – APAB, France

Dragos Gheorghiu (gheorghiu_dragos@yahoo.com),
Department of Research, National University of Arts, Bucharest, Romania

Anthony Harding (a.f.harding@exeter.ac.uk),
University of Exeter, England

Ann-Marie Kroll-Lerner (krollann@pilot.msu.edu),
Michigan State University, Department of Anthropology, USA

George Nash (postmaster@georgenash.demon.co.uk),
Department of Archaeology & Anthropology, University of Bristol, England

Ulla Odgaard (odgaard@archaeology.dk),
Sila – The Greenland Research Centre at the National Museum of Denmark,
Denmark

Paula Purhonen (paula.purhonen@nba.fi),
National Board of Antiquity, Department of Archaeology, Finland

Ralph Rowlett (RowlettR@missouri.edu),
University of Missouri-Columbia, Department of Anthropology, USA

Raymond Thörn (raimond.thorn@malmo.se),
Malmö Museum, Sweden

ARCHAEOLINGUA

Edited by
ERZSÉBET JEREM and WOLFGANG MEID

Main Series

1. **Cultural and Landscape Changes in South-East Hungary. I: Reports on the Gyomaendrőd Project.** Edited by Sándor Bökönyi. 1992. 384 pp. € 36.-. ISBN 963 7391 60 6.

2. Stefan Schumacher: **Die rätischen Inschriften. Geschichte und heutiger Stand der Forschung.** 1992. 2. vermehrte Auflage 2004. 375 pp. € 62.-. ISBN 963 8046 53 8.

3. **Onomasticon Provinciarum Europae Latinarum.** Herausgegeben von Barnabás Lőrincz und Ferenc Redő. Vol. I. 1994. XIV, 364 pp. € 50.-. ISBN 963 8046 01 5 Ö. ISBN 963 8046 02 3 K.

4. **Die Indogermanen und das Pferd. Akten des Internationalen interdisziplinären Kolloquiums, Freie Universität Berlin, 1.-3. Juli 1992. Bernfried Schlerath zum 70. Geburtstag gewidmet.** Herausgegeben von Bernhard Hänsel und Stefan Zimmer. 1994. 272 pp. € 72.-. ISBN 963 8046 03 1.

5. **Cultural and Landscape Changes in South-East Hungary. II.** Edited by Sándor Bökönyi. 1996. 453 pp. € 36.-. ISBN 963 8046 04 X.

6. Garrett S. Olmsted: **The Gods of the Celts and the Indo-Europeans.** 1994. XVI, 493 pp. € 98.-. ISBN 963 8046 07 4.

7. **Die Osthallstattkultur. Akten des Internationalen Symposiums, Sopron, 10. – 14. Mai 1994.** Herausgegeben von Erzsébet Jerem und Andreas Lippert. 1996. 588 pp. € 88.-. ISBN 963 8046 10 4.

8. **Man and the Animal World. Studies in Archaeozoology, Archaeology, Anthropology and Palaeolinguistics in memoriam Sándor Bökönyi.** Edited by Peter Anreiter, László Bartosiewicz, Erzsébet Jerem and Wolfgang Meid. 1998. 720 pp. € 92.-. ISBN 963 8046 15 5.

9. **Archaeology of the Bronze and Iron Age – Environmental Archaeology, Experimental Archaeology, Archaeological Parks. Proceedings of the International Archaeological Conference, Százhalombatta, 3–7 October, 1996.** Edited by Erzsébet Jerem and Ildikó Poroszlai. 1999. 488 pp. € 68.-. ISBN 963 8046 25 2.

10. **Studia Celtica et Indogermanica. Festschrift für Wolfgang Meid.** Herausgegeben von Peter Anreiter und Erzsébet Jerem. 1999. 572 pp. € 78.-. ISBN 963 8046 28 7.

11. **From the Mesolithic to the Neolithic. Proceedings of the International Archaeological Conference held in the Damjanich Museum of Szolnok, September 22–27, 1996.** Edited by Róbert Kertész and János Makkay. 2001. 461 pp. € 72.-. ISBN 963 8046 35 X.

12. Garrett Olmsted: **Celtic Art in Transition during the First Century BC. An Examination of the Creations of Mint Masters and Metal Smiths, and an Analysis of Stylistic Development during the Phase between La Tène and Provincial Roman.** 2001. 340 pp., with 142 plates. € 72.-. ISBN 963 8046 37 6.

13. **The Archaeology of Cult and Religion.** Edited by Peter F. Biehl and François Bertemes with Harald Meller. 2001. 288 pp. € 68.-. ISBN 963 8046 38 4.

14. Gerhard Tomedi: **Das hallstattzeitliche Gräberfeld von Frög. Die Altgrabungen von 1883 bis 1892.** 2002. 706 S., mit 118 Karten, € 76.- ISBN 963 8046 42 2

15. **Morgenrot der Kulturen. Frühe Etappen der Menschheitsgeschichte in Mittel- und Südosteuropa. Festschrift für Nándor Kalicz zum 75. Geburtstag.** Herausgegeben von Erzsébet Jerem und Pál Raczky. 2003. 570 pp. € 78.-. ISBN 963 8046 46 5.

16. **The Geohistory of Bátorliget Marshland.** Edited by Pál Sümegi and Sándor Gulyás. 2004. 360 pp. € 66.-. ISBN 963 8046 47 3.

17. **Nord-Süd, Ost-West. Kontakte während der Eisenzeit in Europa. Akten der Internationalen Tagungen der AG Eisenzeit in Hamburg und Sopron 2002.** Herausgegeben von Erzsébet Jerem, Martin Schönfelder und Günther Wieland. 2006. ca. 320 pp. € 84.-. ISBN 963 8046 57 0.

18. Raimund Karl: **Altkeltische Sozialstrukturen.** 2006. 609 pp. € 78.-. ISBN 963 8046 69 4.

19. Martin Hannes Graf: **Schaf und Ziege im frühgeschichtlichen Mitteleuropa. Sprach- und kulturgeschichtliche Studien.** 2006. 320 pp. € 60.-. ISBN 963 8046 70 8.

20. **Anthropology of the Indo-European World and Material Culture. Proceedings of the 5th International Colloquium of Anthropology of the Indo-European World and Comparative Mythology.** Edited by Marco V. García Quintela, Francisco J. González García and Felipe Criado Boado. 2006. 368 pp. € 64.-. ISBN 963 8046 72 4.

Series Minor

3. Sándor Bökönyi: **Pferdedomestikation, Haustierhaltung und Ernährung. Archäozoologische Beiträge zu historisch-ethnologischen Problemen.** 1993. 61 pp. € 18.-. ISBN 963 7391 65 7.

4. Ferenc Gyulai: **Environment and Agriculture in Bronze Age Hungary.** 1993. 59 pp. € 18.-. ISBN 963 7391 66 5.

5. Wolfgang Meid: **Celtiberian Inscriptions.** 1994. 62 pp. € 20.-. ISBN 963 8046 08 2

6. Marija Gimbutas: **Das Ende Alteuropas. Der Einfall von Steppennomaden aus Südrußland und die Indogermanisierung Mitteleuropas.** 1994. 2. Aufl. 2000. 135 pp. € 32.-. ISBN 963 8046 09 0.

7. Eszter Bánffy: **Cult Objects of the Neolithic Lengyel Culture. Connections and Interpretation.** 1997. 131 pp. € 26.-. ISBN 963 8046 16 3.

8. Wolfgang Meid: **Die keltischen Sprachen und Literaturen. Ein Überblick.** 1997. 2. Aufl. 2005. 94 pp. € 20.-. ISBN 963 8046 65 1.

9. Peter Anreiter: **Breonen, Genaunen und Fokunaten. Vorrömisches Namengut in den Tiroler Alpen.** 1997. 173 pp. € 30.-. ISBN 963 8046 18 X.

10. Nándor Kalicz: **Figürliche Kunst und bemalte Keramik aus dem Neolithikum Westungarns.** 1998. 156 pp. € 30.-. ISBN 963 8046 19 8.

11. **Transhumant Pastoralism in Southern Europe. Recent Perspectives from Archaeology, History and Ethnology.** Edited by Haskel J. Greenfield and László Bartosiewicz. 1999. 245 pp. € 36.-. ISBN 963 8046 11 2.

12. Francisco Marco Simón: **Die Religion im keltischen Hispanien.** 1998. 168 pp. € 32.-. ISBN 963 8046 24 4.

13. Peter Raulwing: **Horses, Chariots and Indo-Europeans. Problems of Chariotry Research from the Viewpoint of Indo-European Linguistics.** 2000. 210 pp. € 36.-. ISBN 963 8046 26 0.
14. John Chapman: **Tension at Funerals – Micro-Tradition Analysis in Later Hungarian Prehistory.** 2000. 184 pp. € 32.-. ISBN 963 8046 29 5.
15. Eszter Bánffy: **A Unique Prehistoric Figurine of the Near East.** 2001. 106 pp. € 24.-. ISBN 963 8046 36 8.
16. Peter Anreiter: **Die vorrömischen Namen Pannoniens.** 2001. 316 pp. € 36.-. ISBN 963 8046 39 2.
17. Paul Gaechter: **Die Gedächtniskultur in Irland.** 2003. 116 pp. € 20.-. ISBN 963 8046 45 7.
18. **The Geoarchaeology of River Valleys.** Edited by Halina Dobrzańska, Erzsébet Jerem and Tomasz Kalicki. 2004. 214 pp. € 38.-. ISBN 963 8046 48 1.
19. Karin Stüber: **Schmied und Frau. Studien zur gallischen Epigraphik und Onomastik.** 2004. 125 pp. € 24.-. ISBN 963 8046 55 4.
20. Wolfgang Meid: **Keltische Personennamen in Pannonien.** 2005. 350 pp. € 40.-. ISBN 963 8046 56 2.
21. **The Archaeology of Cult and Death. Proceedings of the Session "The Archaeology of Cult and Death" Organized for the 9th Annual Meeting of the European Association of Archaeologists, 11th September 2003, St. Petersburg, Russia.** Edited by Mercourios Georgiadis and Chrysanthi Gallou. 2006. 194 pp. € 32.-. ISBN 963 8046 67 8.
22. **Landscape Ideologies.** Edited by Thomas Meier. 2006. 260 pp. € 34.-. ISBN 963 8046 71 6.
23. **The Archaeology of Fire. Understanding Fire as Material Culture.** Edited by Dragos Gheorghiu and George Nash. 2007. 261 pp. € 34.-. ISBN 978-963-8046-79-6.

Eastern Alpine Iron Age Studies

1. **Die Kelten in den Alpen und an der Donau. Akten des Internationalen Symposions St. Pölten, 14.–18. Oktober 1992.** Herausgegeben von Erzsébet Jerem, Alexandra Krenn-Leeb, Johannes-Wolfgang Neugebauer und Otto H. Urban. 1996. Second, revised ed. 2004. 462 pp. € 62.-. ISBN 963 8046 21 X.

Praehistoria

1. **Praehistoria. International prehistory journal of the University of Miskolc.** Edited by Árpád Ringer, Zsolt Mester and Erzsébet Jerem. **Volume 1, 2000.** 188 pp. € 40.-. **Volume 2, 2001.** 201 pp. € 34.-. **Volume 3, 2002.** 338 pp. € 40.-. **Volume 4–5, 2003–2004.** 247 pp. € 40.-. HU ISSN 1586 7811.

Studia Aegyptiaca

1. **The Mortuary Monument of Djehutymes (TT 32)** Vols. I–II. Edited by László Kákosy, Tamás A.Bács, Zoltán Bartos, Zoltán I. Fábián and Ernő Gaál. 2004. 384 pp. € 124.-. Vol. I.: ISBN 963 8046 51 1; Vol. II.: ISBN 963 8046 52 X.

Please address orders to:

ARCHAEOLINGUA

H-1250 Budapest, Pf. 41. Fax: (+361) 3758939

e-mail: kovacsr@archaeolingua.hu http://www.archaeolingua.hu/